Build Your
Own Low-Cost
Signal Generator

Build Your Own Low-Cost Signal Generator

Delton T. Horn

TAB Books

Division of McGraw-Hill, Inc.

New York   San Francisco   Washington, D.C.   Auckland   Bogotá
Caracas   Lisbon   London   Madrid   Mexico City   Milan
Montreal   New Delhi   San Juan   Singapore
Sydney   Tokyo   Toronto

© 1994 by **TAB Books.**
Published by TAB Books, a division of McGraw-Hill, Inc.

pbk 1 2 3 4 5 6 7 8 9 DOH/DOH 9 9 8 7 6 5 4
hc   1 2 3 4 5 6 7 8 9 DOH/DOH 9 9 8 7 6 5 4

Library of Congress Cataloging-in-Publication Data
Horn, Delton T.
    Build your own low-cost signal generator / by Delton T. Horn.
        p.   cm.
    Includes index.
    ISBN 0-07-030428-9   ISBN 0-07-030429-7 (pbk.)
    1. Signal generators—Design and construction—Amateurs' manuals.
    I. Title.
    TK7872.S5H67   1994
    621.3815'48—dc20   94-14281
                                                            CIP

Acquisitions editor: Roland S. Phelps
Editorial team: Joanne Slike, Executive Editor
                Andrew Yoder, Managing Editor
                Anita Louise McCormick, Book Editor
Production team: Katherine G. Brown, Director
                Susan E. Hansford, Coding
                Jana L. Fisher, Desktop Operator
                Kelly S. Christman, Proofreading
                Joann Woy, Indexer
Design team: Jaclyn J. Boone, Designer
                Brian Allison, Associate Designer
Cover photograph: David G. Colwell, Smithsburg, Md.
Cover design: Graphics Plus, Littlestown, Pa.   EL1
Cover copy writer: Cathy Mentzer   0304297

Contents

Projects

Introduction

Few electronics systems would be of much use without signals of some sort flowing through them. Sometimes the signal is an electrified version of some non-electronic parameter. For example, a PA amplifier system uses voice and other sound signals converted into electrical form by a microphone. An electronic light meter's signal is the electrical equivalent of the intensity of the light shining on some sort of photosensor.

Often, though, the signals needed are purely electronic in origin. The circuitry must create its own signals from scratch. There are countless types of electrical signals that we might want to generate. Simply put, an electronically-generated signal may be either a dc voltage or current, or an ac waveform of some type. Both dc and ac signal generation are covered thoroughly in this book. Chapter 1 deals with dc signals, and the rest of the book explores various ac signals, including the most popular and widely used—the sine wave (chapter 2), the rectangle wave and its special forms, the square wave and the pulse wave (chapter 3), the triangle wave, the sawtooth wave, and the staircase wave.

Odd and exotic special purpose waveforms are also covered, along with white noise and pink noise. Function generators, which can generate two or more different waveforms within a single circuit, are discussed in chapter 4. White-noise and pink-noise generators are the subjects of chapter 6.

Chapter 7 explores various ways to generate nonstandard, complex ac waveforms. Amplitude modulation, frequency modulation, additive synthesis and subtractive synthesis are among the techniques discussed here. The special problems and considerations involved when working with RF (radio frequency) signals are explored in chapter 7.

Finally, chapter 8 examines the various, and often surprising ways that ac signals can be generated by digital circuitry. Usually digital circuitry only works with rectangle waves of some sort, but techniques for digitally synthesizing other analog waveforms will also be considered in this chapter.

In addition to the theoretical background and circuit design tips, this book also features sixteen inexpensive but useful projects, including sound generators, power supplies, and test equipment. Certainly, none of these projects will put any commercial manufacturers of such equipment out of business. For the most part, we won't be dealing with super-precision specifications in any of these projects. But the specs are still surprisingly good for such easy, minimal-cost projects. They should be good enough for most typical hobbyist applications. The limitations of any given project will be mentioned in the text, where relevant.

You can spend hundreds of dollars to buy a deluxe commercial signal generator. Or, you can build any one of these projects for less than $25, and it will do almost as good of a job for most noncritical applications.

With this book as your guide, you should be able to generate almost any electronic signal you'll ever need, without having to spend a bundle.

1
CHAPTER

dc power supplies

A *dc voltage* usually isn't thought of as a *signal*, even though it is as much of a signal as any ac waveform. It is just a steady-state signal, rather than a fluctuating signal.

dc voltages probably aren't normally considered to be signals because of the way they are used. When used to power a circuit, the dc voltage does not really function as a signal. However, in many circuits, dc voltages are used as control signals. For example, in a PLL or servo circuit, a dc signal is fed back to make automatic self-correction adjustments. In older analog music synthesizers, almost all functional parameters were set by various control voltages, a clear case of dc voltages being used as signals.

We are beginning with dc voltages in this book because they are the simplest possible type of signal. The *waveform* is simply a straight line. The polarity doesn't change, and the voltage value is constant (until the signal is changed to a new value). The frequency is always 0 Hz. All ac signals are necessarily more complex. We'll get to them in later chapters of this book.

In most modern electronics work, two basic sources are used to obtain dc voltages—batteries and ac-to-dc power supplies. *Batteries* are perfectly straight-forward. An internal chemical reaction of some sort generates a dc voltage, which can be used as it is, or it can be dropped to a lower value by a voltage-divider network. (Voltage-multiplier circuits also exist for increasing a dc voltage, at the expense of the available current.) For batteries, dc is the natural order of things.

An *ac-to-dc power supply* is a circuit that accepts ordinary ac house current, and converts it to a specific dc voltage. Some sort of low-pass filtering is almost always used to reduce the amount of ac fluctuations sneaking through the circuit to the nominally dc output voltage. Such ac fluctuations riding on a dc voltage are known as *ripple* (sometimes *ac ripple*, although this term is redundant), and the effect is illustrated in Fig. 1-1.

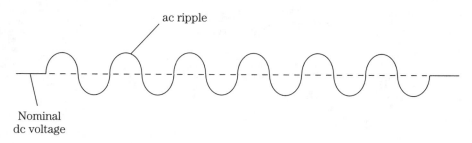

1-1 ac fluctuations riding on a dc voltage are known as ripple.

Better-quality power-supply circuits also include some form of voltage regulation, which reduces ripple effects even more than simple filtering, while minimizing the effects of loading. Without voltage regulation, the output voltage can vary (often by a very significant amount) with changes in the current drawn by the load. Such load variations are almost always highly undesirable. In most applications, we definitely want the supply voltage to remain constant, regardless of the load.

The basics of power-supply circuits

A simple power-supply circuit typically consists of three basic elements—a power transformer, a rectifier, and a filter. The *power transformer* drops the input ac voltage (nominally 120 Vac) to a lower ac voltage, closer to (but higher than) the desired output dc voltage. For example, in a power-supply circuit intended to put out 12 volts dc, the transformer might drop the ac voltage down to 13.6 Vac. When used in this way, the transformer is called a *step-down transformer*. In some special cases, a *step-up transformer* can be used to create a voltage larger than the nominal ac input. An *isolation transformer* has an output voltage exactly equal to its input voltage, but the two sections are electrically isolated from each other, maximizing safety, and almost totally eliminating loading effects on the transformer voltage.

After the transformer comes the *rectifier*. This can be a single diode, producing a half-wave rectifier circuit. Multiple diodes can be used to create more sophisticated power-supply circuits known as *full-wave rectifier* and *bridge rectifier* circuits. We will discuss each type of rectification shortly.

Finally, the filter, as already noted, reduces any stray ac content, or ripple, in the output voltage. In most simple power-supply circuits, the *filter* is simply a large-value electrolytic capacitor, which shunts any ac signal content to ground. Better quality power-supply circuits might use more sophisticated filtering, but the basic principle is the same. There is a low-resistance path to ground for ac signals, but this path presents a very high resistance to a dc voltage, so the ac content is shunted off, while the desired dc voltage has nowhere to go but to the circuit's output terminals.

Half-wave rectifiers

A simplified *half-wave rectifier* circuit (without a filter) is shown in Fig. 1-2. The ac voltage from the transformer, by definition, reverses its polarity twice during each cycle. For one half of each cycle, terminal A is positive with respect to terminal

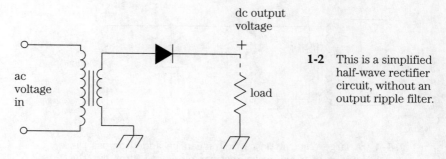

dc output
voltage

1-2 This is a simplified
half-wave rectifier
circuit, without an
output ripple filter.

B (which is grounded here). For the other half of each cycle, terminal A is negative
with respect to terminal B.

A *diode*, or rectifier is a *polarized device*. It only permits a voltage to pass
through it in one direction (polarity). In the other direction (polarity), the voltage's
path through the rectifier is blocked. In our half-wave rectifier circuit, when terminal A is positive, the diode is *forward-biased*, so the applied voltage can pass
through it to the output.

However, when terminal A is negative, the diode is *reverse-biased*, blocking the
applied voltage. Nothing at all appears at the output during the negative half-cycles.
Only the positive half-cycles appear at the circuit's output, as illustrated in Fig. 1-3.

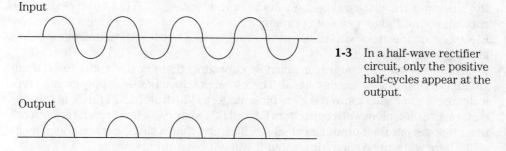

1-3 In a half-wave rectifier
circuit, only the positive
half-cycles appear at the
output.

Of course, this is not a true dc voltage at all. Half the time, there is no output
voltage at all. The rest of the time, the output voltage is either in the process of rising from zero to the maximum level, or dropping from the maximum level back down
to zero. It never holds a constant value.

To achieve a closer approximation of a true dc voltage, we need to add a filter
stage to our half-wave rectifier circuit. This is usually accomplished by placing a
large-valued capacitor across the diode's output, as shown in Fig. 1-4.

Let's start things out at zero. The ac signal is in a negative half-cycle, so the output voltage is zero for this instant. We will assume that the capacitor is fully discharged (charge = 0). Nothing will happen until a positive half-cycle begins, and the
diode starts to conduct, permitting a voltage to appear at the output. As the output
voltage rises from zero to its peak value, the capacitor is charged. When the voltage
drops off from its maximum level, the capacitor starts to discharge through the load,
so the voltage that is seen by the load looks something like Fig. 1-5. If the capacitor

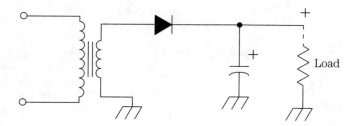

1-4 A large electrolytic capacitor can be added across the output of a half-wave rectifier circuit to filter out the worst of the ripple.

1-5 Adding a filter stage to a half-wave rectifier circuit will give a closer approximation of a true dc voltage.

is large enough, it will not fully discharge before the next positive half-cycle starts. In other words, the capacitor will be repeatedly charged and partially discharged. The less it is discharged when the new cycle begins, the less ripple there will be in the output signal.

The larger the capacitance value, the slower the discharge rate, and therefore, the shallower the discharging angle in the output waveform. That is, increasing the capacitance will give a closer approximation of a true dc voltage with less ripple. Electrolytic capacitors with values of several hundred to a few thousand microfarads are typically used.

But even with the largest imaginable capacitor, there is still going to be a fair amount of ripple in the output signal. Therefore, practical power-supply circuits typically use a somewhat more complex filter stage, as illustrated in Fig. 1-6. In this circuit, resistor R2, along with capacitors C1 and C2, comprise a low-pass filter network that smooths out the output signal more efficiently than a single capacitor can by itself. There will still be some ripple, but it will not be as pronounced.

Resistor R1 is a *surge resistor* that protects the diode from any sudden increase in the current drawn through the circuit. The surge resistor typically has a fairly small value, so normally the voltage drop across it is negligible. But an increase in the current drawn through the surge resistor will cause its voltage drop to increase pro-

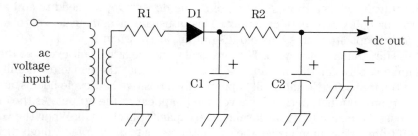

1-6 Most practical power-supply circuits typically use a somewhat more complex filter stage.

portionately, since according to Ohm's law, voltage equals current times resistance ($E = IR$) and the resistance is a constant in this case. Sometimes, surge resistor R1 is also fused for additional protection.

A surge resistor has uses other than for protection against unusual or abnormal circuit defects. In some circuits, it is also needed for normal operating conditions. Assume that no source (ac) voltage at all is being applied to the circuit. Any residual charge on the capacitors will soon be fully discharged through resistor R2 and the load circuit. The capacitors are now completely discharged. Now, when power is first applied, the capacitors will tend to draw a large amount of current until they are almost completely charged. Assuming the capacitance values are large enough, there won't be sufficient time for the capacitors to fully charge during a single cycle, so it takes a few cycles for ordinary operation to begin. During this time, more current will be drawn through the diode. Of course, this extra current drain will increase the voltage drops across the other components, and again, the surge resistor protects the diode from burning itself out by attempting to conduct more current than it can safely handle.

In many better-quality power-supply circuits, a *thermistor* (temperature-sensitive resistor) is used for the surge resistor. When power is first applied to the circuit, the components, including the thermistor tend to be relatively cool. The thermistor has a higher resistance when it is cold. This means that when power is first applied, there is a relatively large voltage drop across the thermistor, leaving only a relatively small voltage to pass through the diode. As current passes through the circuit, the components start to dissipate heat. The increased temperature causes the resistance of the thermistor (and thus, the voltage drop across it) to drop to a fairly low value. From then on, it acts like any ordinary (fixed-value) surge resistor.

A half-wave rectifier power-supply circuit is simple and inexpensive, but it leaves much to be desired. Even with the best possible filtering, the ripple content of the output voltage will inevitably be relatively high. This type of circuit is also energy-wasteful. Half of each input cycle is completely unused. This energy is simply dissipated as heat, and does no good, even though it adds to the input power consumed by the circuit. Clearly, a more efficient type of power-supply circuit is highly desirable in many, if not most practical applications—especially if relatively large power levels are required.

Full-wave rectifiers

By using two diodes in parallel with opposing polarities, we can create a more efficient type of power-supply circuit. Since this type of circuit uses both halves of the input cycles, it is known as a *full-wave rectifier*.

A simplified full-wave rectifier circuit (without any filter stage) is shown in Fig. 1-7. Notice that a full-wave rectifier circuit must always be used with a center-tapped transformer, with the center-tap grounded.

Remember, if the center-tap of a transformer's secondary winding is grounded, the lower half of the secondary winding will carry a signal that is equal to, but 180 degrees out-of-phase with, the signal carried by the upper half of the secondary winding. This means that in our full-wave rectifier circuit, when diode D1 is passing a positive half-cycle, diode D2 is blocking a negative half-cycle. And similarly, when diode D1 is blocking a negative half-cycle, diode D2 is passing a positive half-cycle.

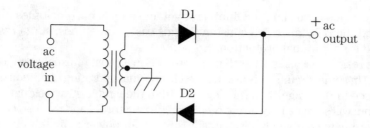

1-7 This is a simplified full-wave rectifier circuit, without an output ripple filter.

One of the diodes is conducting, and the other is non-conducting at all times. This means the output signal will resemble Fig. 1-8. An actual, non-zero voltage will be present at virtually all times, except for those brief instants when the original waveform crosses through the 0 volts line, in either direction.

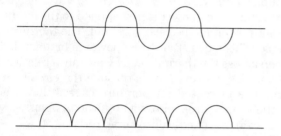

1-8 In a full-wave rectifier circuit, one of the diodes is conducting, and the other is non-conducting at all times.

Besides wasting less input power, the output signal of a full-wave rectifier circuit is easier to filter, because there is less time for the filter capacitor to discharge before it is charged again. The simplest type of filtering for a full-wave rectifier signal, and the resulting output signal, are illustrated in Fig. 1-9. Notice that both the positive (D1) and the negative (D2) output lines need their own filter capacitor, and they are both isolated from the ac ground.

The chief disadvantage of the full-wave rectifier circuit is the requirement for a center-tapped transformer, which is usually more expensive than a non-center-tapped transformer.

Bridge rectifiers

A *bridge rectifier* circuit, like the one shown in Fig. 1-10, combines the advantages of both a half-wave rectifier and a full-wave rectifier. Like the full-wave rectifier, the bridge rectifier uses the entire input cycle, and its output signal is fairly easy to filter.

On the other hand, like the half-wave rectifier, the bridge rectifier does not require an expensive center-tapped transformer as is necessary with the full-wave rectifier. While a bridge rectifier requires four diodes (instead of one for a half-wave rectifier, or two for a full-wave rectifier), it is still usually more economical for semiconductor circuits than a full-wave rectifier, in which the center-tapped transformer is usually the greatest expense.

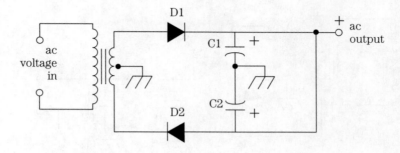

1-9 This is the simplest type of filtering for a full-wave rectifier circuit, and the resulting output signal.

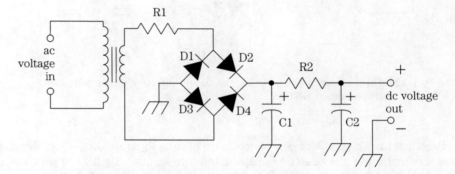

1-10 A bridge-rectifier circuit combines the advantages of both a half-wave circuit rectifier and a full-wave rectifier.

The bridge rectifier circuit also requires a bit less space, and produces less heat. (Bridge rectifier circuits using tube diodes are not practical.) Another potentially helpful way in which a bridge rectifier circuit resembles a half-wave rectifier circuit is that one of the output lines can be at true ground potential.

The operation of a bridge rectifier is not as obvious and straight-forward as either a half-wave rectifier or full-wave rectifier circuit. It is easiest to understand if we re-draw the circuit diagram for each half-cycle, showing only the forward-biased (conducting) diodes. At any point of the input cycle, two of the diodes in the bridge are conducting and two are reverse-biased. For the positive half-cycles, the circuit effectively acts like the modified circuit shown in Fig. 1-11. The equivalent circuit for the negative half-cycles is illustrated in Fig. 1-12.

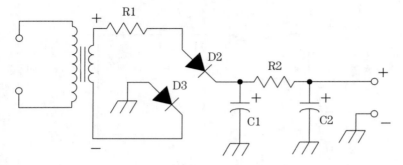

1-11   On the positive half-cycles, the circuit effectively functions like this modified circuit.

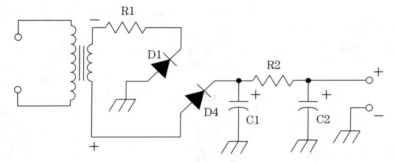

1-12   On the negative half-cycles, the circuit effectively functions like this modified circuit.

Practical bridge rectifiers can be made up of four separate diodes, as shown in these schematics, or they can be encapsulated into a single, dedicated package, as shown in Fig. 1-13. This is usually done simply to conserve space. In some cases, using a dedicated bridge rectifier unit can lower the total circuit cost slightly, but there is rarely a significant difference.

Even though a dedicated bridge-rectifier package looks very different, electrically, it is the exact equivalent to four discrete diodes. There is no functional difference. The circuitry isn't going to care which you use.

When four separate diodes are used to build a bridge rectifier, they must all be closely matched. In other words, you should use the same type number for all four diodes in the bridge. If the diodes have different operating circuits, the bridge rectifier circuit will be thrown out of balance, and it either won't work, or the output voltage will be too far off from the intended value. Of course, when working with any pre-packaged dedicated bridge rectifier unit, you can count on all of the internal diodes being matched with reasonable precision.

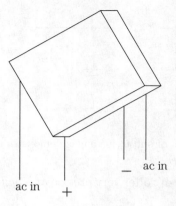

ac in

ac in

+

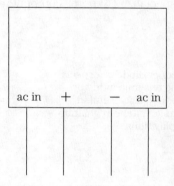

ac in   +   —   ac in

1-13 Some practical bridge rectifiers are encapsulated into a single, dedicated package.

Voltage regulation

One problem with all of the power-supply circuits we have discussed so far is that the output voltage is dependent, to a large extent, on the amount of current drawn by the load circuit. If the current drawn by the load increases for any reason, the voltage drop across the components in the power-supply circuit itself also rises. This inevitably results in a lower output voltage from the power-supply circuit. Of course, just the opposite can happen if the current drawn by the load circuit decreases for any reason—the power supply's output voltage will increase.

One partial solution is shown in the half-wave rectifier circuit of Fig. 1-14. In this circuit, what would ordinarily be resistor R2 is replaced with a type of coil known as a *choke*. This choke coil will tend to oppose any change in the voltage passing through it. Besides the simple voltage regulation, the voltage change resistance effect of this coil results in better ripple filtering than would be possible if an ordinary resistor were used. Since the dc resistance of a coil is extremely low, there is very little waste power from voltage dropped across the choke. Using a choke for voltage

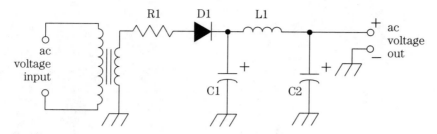

1-14 One partial solution to the problem of output loading is demonstrated in this half-wave rectifier circuit.

regulation is better than nothing, but it doesn't offer very precise or efficient results. A zener diode can do a much better job.

A *zener diode* is a special type of semiconductor diode. A somewhat modified schematic symbol is used to indicate this specialized type of component, as illustrated in Fig. 1-15.

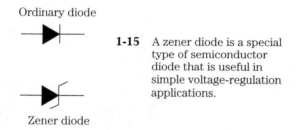

1-15 A zener diode is a special type of semiconductor diode that is useful in simple voltage-regulation applications.

Any ordinary diode blocks current flow (presents a very high resistance) when it is reverse-biased, but conducts heavily (presents a very low resistance) when it is forward-biased. The action of an ordinary diode is shown in graph form in Fig. 1-16. Of course, no practical component is perfect. If the reverse-bias voltage is made large enough (exceeding the *peak inverse voltage* [PIV] rating of the diode) the pn junction will be shorted, and the diode will then conduct, even though reverse-biased. However, the diode will be ruined in the process. The damage is permanent.

A zener diode behaves much like an ordinary diode when forward-biased. When it is reverse-biased, it blocks current flow if the applied voltage is fairly low. If the applied voltage exceeds a specific *avalanche voltage*, the zener diode conducts heavily, but without damaging the zener diode itself. A graph of the action of a zener diode is shown in Fig. 1-17.

A zener diode does have a PIV value that must never be exceeded—or the component will be permanently damaged, just like an ordinary diode. But this PIV limit is much higher than the avalanche voltage, and it is usually a major consideration in most practical zener diode circuits.

To understand how a zener diode is used for voltage regulation, consider the simple circuit shown in Fig. 1-18. We will assume that the voltmeter has an infinite input impedance, so it has no loading effects. (This is for convenience of discussion, any real voltmeter will have a finite input impedance, of course.)

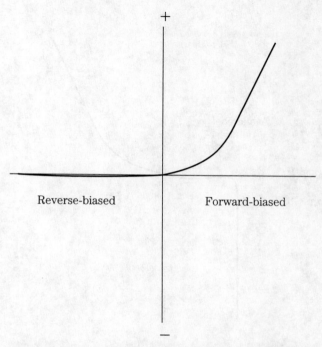

Reverse-biased   Forward-biased

1-16 An ordinary diode tends to block current flow when it is reverse-biased, but conducts heavily when it is forward-biased.

R1 and R2 form a simple voltage divider network. Let's assume that R1 has a value of 1 kΩ (1,000 ohms), and R2 is a 50 kΩ (50,000 ohms) potentiometer. Let's say the input voltage to this circuit is a constant 10 volts dc. The amount of voltage read on the voltmeter will depend on the setting of potentiometer R2.

For example, let's say R2 is set for 1 kΩ (1,000 ohms). R1 and R2 are in series, so the total resistance in the circuit is equal to:

$$R_t = R_1 + R_2$$
$$= 1000 + 1000$$
$$= 2000 \text{ ohms}$$

Using Ohm's law, we can find out how much current is flowing through the circuit:

$$I = \frac{E}{R_t}$$

$$= \frac{10}{2000}$$

$$= 0.005 \text{ ampere}$$

$$= 5 \text{ mA}$$

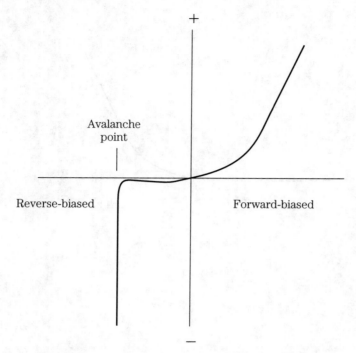

1-17 When a zener diode is reverse-biased, it blocks current flow only if the applied voltage is fairly low, but higher reverse-bias voltages cause it to conduct very heavily.

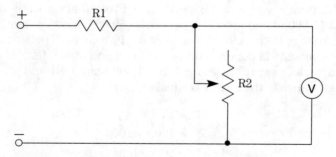

1-18 This simple circuit illustrates the problem of output loading and the need for voltage regulation.

The same current will flow through each series-connected resistor. Now, we can use Ohm's law again to find the voltage drop across R2, which is monitored by the voltmeter:

$$E_2 = IR_2$$
$$= 0.005 \times 1000$$
$$= 5 \text{ volts}$$

Now, let's see what happens when potentiometer R2 is readjusted for a value of 10 kΩ (10,000 ohms). First, the total series resistance in the circuit now becomes:

$$R_t = 1000 + 10000$$
$$= 11,000 \text{ ohms}$$

So the current flow is equal to:

$$I = \frac{10}{11000}$$

$$= 0.00091 \text{ ampere}$$
$$= 0.91 \text{ mA}$$

And the voltage drop across R2, as indicated by the voltmeter, is now equal to:

$$E_2 = 0.00091 \times 10000$$
$$= 9.1 \text{ volts}$$

Quite a difference. Several other examples are summarized in Table 1-1.

What does this have to do with anything? Assume R1 is the total internal resistance of the power-supply circuit, and R2 is load resistance of the circuit powered by the power supply. The effective load resistance changes with any variations in the current drawn by the load circuit, according to Ohm's law. So this simple circuit simply demonstrates the problem of variable loading, and why voltage regulation is so often necessary.

**Table 1-1. Typical effects of loading in
the simple demonstration circuit of Fig. 1-18**

E = 10 volts
R_1 = 1,000 ohms

R_2	Total resistance	I	Output voltage
500	1,500	0.00667	3.33
1000	2,000	0.00500	5.00
2000	3,000	0.00333	6.67
3500	4,500	0.00222	7.78
5000	6,000	0.00167	8.33
6900	7,900	0.00127	8.73
7700	8,700	0.00115	8.85
10000	11,000	0.00091	9.09
12000	13,000	0.00077	9.23
17500	18,500	0.00054	9.46
25000	26,000	0.00038	9.62
33300	34,300	0.00029	9.71
50000	51,000	0.00020	9.80

Now, let's put a zener diode in parallel with the load resistance (R2), as illustrated in Fig. 1-19. We will assume this zener diode has an avalanche voltage of 6.8

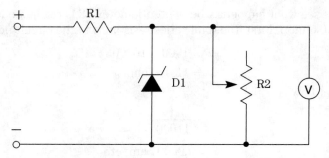

1-19   The addition of a zener diode regulates the output
voltage from the simple circuit of Fig. 1-18.

volts. The output voltage for various load resistances (settings of R2) are summarized in Table 1-2. Notice that for voltages below the critical value (6.8 volts), there
is no difference from the ordinary, unregulated circuit. But the output voltage is
never permitted to exceed 6.8 volts, even with further increases in the load resistance (R2). The simple addition of the zener diode to the circuit has regulated the
output voltage—at least against over-voltages. Fluctuations in the current drawn by
the load are shunted to ground through the zener diode, so they do not affect the
output of the power supply.

**Table 1-2. Typical effects of loading in the simple
voltage-regulated demonstration circuit of Fig. 1-19**

$$E = 10 \text{ volts}$$
$$R_1 = 1,000 \text{ ohms}$$
$$V_z = 6.8 \text{ volts}$$

R_2	Total resistance	I	Output voltage
500	1,500	0.00667	3.33
1000	2,000	0.00500	5.00
2000	3,000	0.00333	6.67
3500	4,500	0.00222	6.80
5000	6,000	0.00167	6.80
6900	7,900	0.00127	6.80
7700	8,700	0.00115	6.80
10000	11,000	0.00091	6.80
12000	13,000	0.00077	6.80
17500	18,500	0.00054	6.80
25000	26,000	0.00038	6.80
33300	34,300	0.00029	6.80
50000	51,000	0.00020	6.80

Voltage regulation using zener diodes is inexpensive, and fairly effective. However, in more critical applications, better voltage regulation might be desirable, or
even essential. In such cases, a special voltage-regulator circuit will be used. A simplified basic voltage-regulator circuit is illustrated in Fig. 1-20. This is just one typical circuit. A number of variations are possible, and are frequently encountered in
practical electronics work.

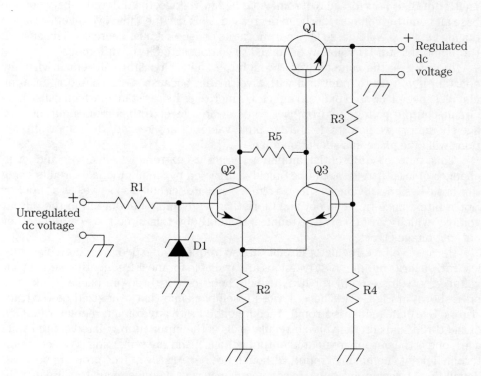

1-20   This is a simplified basic voltage-regulator circuit.

Notice that this circuit involves a lot more than our previous crude voltage regulators. In this particular circuit, a zener diode is used along with three npn transistors. Q1 is used as a *pass transistor*, Q2 and Q3 comprise a difference amplifier that
is set up to detect any error in the output voltage. The zener diode sets up the reference voltage the output voltage will be continuously compared to, but the load on
the reference voltage is constant and low. The two resistors (R3 and R4) between
the output and ground (with the base of transistor Q3 tapped in between them) form
a voltage divider.

If the output voltage increases even slightly because of a decrease in the current
drawn by the load circuit, the voltage on the base of transistor Q3 will be increased
proportionately. Meanwhile, the voltage on the base of transistor Q2 is held constant

by the zener diode reference voltage. Ordinarily, these two base voltages should be equal, keeping the circuit in balance. But we now have a condition where Q3's base is at a higher voltage level than Q2's base. This is the difference detected by the differential amplifier circuitry.

The increase in base voltage on Q3 proportionately increases the emitter and collector currents of this transistor. This means the voltage drop across the shared emitter resistor (R2) must also increase in accordance with Ohm's law ($E = IR$). Since the emitter of Q2 is now forced to a somewhat higher voltage, the difference between the base and emitter voltages has been decreased. This has the same overall effect as reducing Q2's base voltage, so Q2's output (at its collector) must decrease. The current through transistor Q1 will now be forced to compensate for the difference.

In practice, the voltage drop through the emitter resistor will remain virtually constant if you try to monitor it with a voltmeter, because these self-compensating changes take place so rapidly. In effect, Q1 increases its resistance, which causes the circuit's output voltage to drop back to its desired level. If the circuit's output voltage should drop below its desired nominal value for any reason, the opposite reactions will take place in a similar manner.

This type of differential amplifier circuit is extremely sensitive, and even changes of less than 0.1% can be quickly corrected by some practical circuits using this basic design. Obviously, such a voltage-regulator circuit also serves as a superior ripple filter, since ripple is nothing more than a rhythmic fluctuation in the output voltage, which can be corrected almost instantly by this circuit, as can any other output voltage error.

Because voltage-regulator circuits are so frequently needed in modern electronics work, a large number of IC versions are available for most frequently-used output voltages. A voltage-regulator IC is usually housed in a three-pin package, like the ones shown in Fig. 1-21. These devices resemble somewhat over-sized power transistors. Use of heatsinks is strongly recommended with any voltage-regulator IC. One of the three leads on the voltage-regulator IC is the input (unregulated source voltage), one is the output (regulated output voltage), and the remaining is the common terminal point (nominally ground, although not necessarily at true ground potential, depending on the application). In schematic diagrams, voltage regulators are usually drawn as simple boxes, as shown in Fig. 1-22.

The most commonly available voltages for voltage-regulator ICs are 5, 12, and 15 volts. Other voltages are also available, although they might be more difficult to locate. Most voltage-regulator chips are designed for use in either positive or negative

1-21 A voltage-regulator IC is usually housed in a three-pin package, somewhat resembling over-sized power transistors.

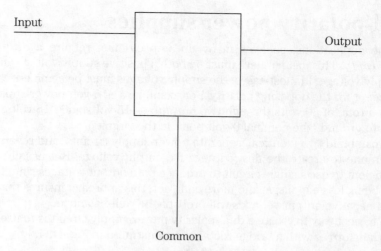

1-22 In schematic diagrams, voltage regulators are usually shown as simple boxes.

ground operation—that is, the output voltage can be either negative or positive. In most cases, these two types are not interchangeable. You must use a voltage-regulator IC designed for the desired output voltage polarity. The two types of voltage-regulator ICs can be used together to create a dual-polarity power supply, as illustrated in Fig. 1-23.

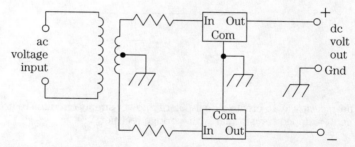

1-23 A positive-voltage regulator can be used together with a negative-voltage regulator in a dual-polarity power-supply circuit.

For most voltage-regulator ICs, the input voltage can vary over a rather large range without affecting the output voltage. For example, one typical 5-volt regulator on the market accepts input voltages of up to 35 volts.

Practical voltage-regulator ICs are limited in the amount of current that can be safely drawn from them. If they are forced to put out too much current, the delicate semiconductor crystals within the IC will be over-heated, and are likely to be damaged or destroyed. If greater current is required than the available type of voltage-regulator IC can handle, several voltage-regulators can be used in parallel.

Dual-polarity power supplies

Some applications, such as many op amp circuits, require a *dual-polarity power supply*. This means there must be both a positive supply voltage and a negative supply voltage. In most cases, the supply voltages must be symmetrical, that is, equal except for the opposing polarity. For example, a ±15-volt power-supply circuit puts out a total of 30 volts through two outputs—+15 volts and –15 volts, both referenced to ground (the nominal 0-volt point in the circuit.)

You can build two identical, separate power-supply circuits, and reverse the polarities in one to create the desired negative output voltage. But usually, it will be more efficient to use a single circuit to produce both output voltages simultaneously. Besides being less expensive and more compact, this approach insures that the positive and negative output voltages will probably be well-balanced.

The easiest way to create a dual-polarity power-supply circuit is to use a center-tapped transformer with a bridge rectifier, as illustrated in Fig. 1-24.

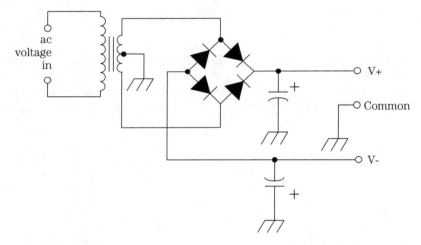

1-24   The easiest way to create a dual-polarity power-supply circuit is to use a center-tapped transformer with a bridge rectifier.

Safety and power supplies

A power-supply circuit, by definition, uses the full house current voltage—nominally about 120 volts ac—and that can be plenty to cause a severe shock, or even death, if not treated with proper respect and care.

For reasons of safety, all conductors carrying ac power must be fully enclosed in a plastic or other non-conducting case before power is applied. It should be impossible for anyone to ever touch a live wire, even if an unanticipated short circuit should occur. If a wire splice or other connection simply must be out in the open, wind it carefully in several layers of electrician's tape, enclose it in heat-shrink tubing, or secure the connection with an appropriate wire nut. Do not leave any wires exposed. Ever.

Always use a fuse of the appropriate size in the ac input line, whether it is shown in the schematic diagram of a project or not. Such protection should always be assumed in every ac-powered electronics project. Don't take foolish chances. The results of saving a few pennies on a fuse could be a fire, and/or serious injury, or even death. If you're very lucky, maybe only some expensive parts in your project will be destroyed.

A *fuse* is just a short length of special wire or a metallic strip that is designed to melt when its temperature exceeds some specific value. This delicate metal strip is normally protected within a small glass tube, or within some other housing. When a current flows through any conductor, that conductor will be heated up in a predictable way. Thus, the temperature is directly proportional to the current flow. So, when the current exceeds the maximum acceptable value, the temperature is sufficient to melt the metal strip or wire of the fuse. This opens up the circuit, and prevents any further current flow.

There are several different types of fuses. Two types which you should be aware of are *slow-blow* (sometimes spelled slo-blow) fuses and *fast-blow*, or regular fuses. A slow-blow fuse will tolerate a transient over-current signal for a brief period of time without blowing. A fast-blow fuse will burn out almost instantly, as soon as the rated current value is exceeded. Each type of fuse is suitable for different applications. For example, slow-blow fuses are the best choice for protecting loudspeakers from excessive signal levels, but permitting brief transients that frequently occur in music, and that don't present a risk to the speakers unless they continue too long. In power-supply circuits, however, you should always use fast-blow fuses. Excessive current can damage or destroy the semiconductor crystal of an expensive IC almost instantly—it's a lot cheaper and more convenient to replace the fuse than the IC.

Occasionally sharp transients might appear on the ac power line, and could cause a fuse to blow, even though nothing is wrong. If a fuse in your equipment blows, disconnect power and replace the fuse with an identical unit. Reconnect the power and turn the equipment back on. If everything works fine now, assume that the culprit was just a stray transient. Don't worry about it. The problem has been fully corrected. The fuse did its job. If the fuse hadn't been there, something in the circuitry might have been burnt out by that transient, possibly leading to an expensive, major repair job.

If the plans for a project specify a fuse of a specific size, that is the size you should use. If you don't know the appropriate rating for a fuse, try to calculate the maximum current the circuit will be required to carry. You don't have to worry about exact precision here, just approximate. Next, add about one quarter to one third of this estimated value, and use the nearest standard value fuse. For example, let's say you expect your project to handle currents up to 0.90 ampere. In this case, a fuse rated anywhere from 1.125 ampere up to 1.20 ampere would be indicated. If you can find a 1⅛-amp (1.125 ampere) fuse, that would be ideal, but this is a rather odd, hard (though not impossible) value to find. A standard 1¼-(1.25 amp) fuse is a little higher than we calculated, but it is still reasonably close. But a 1.5-amp fuse is probably too high, and using it might be asking for trouble.

Do not over-rate the fuse. Don't use an over-rated fuse or bypass a fuse even temporarily for testing purposes. If the fuse is too large, it won't blow in time. Something further along in the circuit (usually an expensive semiconductor component) is likely to blow out to protect the fuse—which is surely a case of false economy. If the correctly-rated fuse repeatedly blows as soon as power is applied, or after only a few minutes of operation, it's almost guaranteed that there is a short circuit somewhere in the current path. Disconnect all power and trace out the problem with an ohmmeter or other test equipment.

Always disconnect the power cord before replacing a fuse. Don't take chances on getting a serious, possibly fatal, electrical shock. The risk isn't worth saving a couple seconds to make sure you are working safely.

Instead of a fuse, you might use a *circuit breaker*. This is a reusable device that is designed to trip and open up a small mechanical switch when the current flow exceeds a specific value. Electrically, the effect is the same as if a fuse was used, but the circuit breaker can be manually reset by pushing a small button. It does not need to be replaced after it pops. Usually there is no need to disconnect power to the circuit to reset the circuit breaker, since the plastic body of the unit and the reset button are fully insulated.

If a circuit breaker repeatedly pops open, the odds are that a short circuit exists somewhere in the current path, as with repeatedly blowing fuses. However, a circuit breaker itself will occasionally develop a defect. This is rather uncommon, but it can happen. If you encounter a circuit breaker that keeps popping open, first assume there is a short circuit, but if you can't find anything wrong with the circuitry protected by the circuit breaker, the easiest way to test it is with another circuit breaker with the same rating, which is known to be good. You can either replace the original circuit breaker with the known good unit, or you can temporarily wire the new circuit breaker in parallel with the original one. (Disconnect all power before attempting either of these changes.)

If the new circuit breaker works correctly, you've solved the problem—the original circuit breaker is defective. Simply replace it permanently. However, if the known good circuit breaker also keeps popping open, there is something wrong in the protected circuit that you have missed. Never bypass any circuit breaker (even temporarily) or replace it with a higher-rated unit. The risk of doing serious damage to your equipment and/or yourself is too great. It is NEVER worth the risk.

Designing a power-supply circuit

So far, we have been looking at power-supply circuits in a very general way. In practical electronics work, you will obviously need to convert this theoretical information into a practical circuit with specific component values to suit the needs of your intended application. Designing a simple power-supply circuit is seldom difficult. Only a handful of components are required, and usually there will be considerable leeway in their values. You usually won't need to work out any calculations too precisely.

In most simple power-supply circuits, the transformer is the most critical (and most expensive) component. There are four basic specifications for a power transformer:

- Primary voltage
- Secondary voltage
- Power rating
- Regulation factor

The *primary voltage* is simply the ac voltage the transformer expects to see at its primary winding, or its input. In the U.S.A., most commonly available transformers are designed for use on ordinary ac house current, which has a nominal value of about 120 Vac. Sometimes the primary transformer is marked 110 Vac or 117 Vac. Different manufacturers use slightly different standards. The actual voltage of ac house current tends to fluctuate, so these are all just nominal values, and they are all considered to be identical.

You might encounter a transformer with a primary voltage rating of 220 Vac or 240 Vac. Can you use such a transformer with ordinary house-current? Sure, but you will get a different secondary voltage than what is marked on the transformer's housing. The primary voltage is basically a reference that determines the secondary voltage. Reducing the primary voltage will result in a lower secondary voltage, and vice versa.

Occasionally you might come across a power transformer with a low primary voltage rating, such as 32 Vac. You might be able to use such a transformer on ordinary house current, but I wouldn't count on it. The excessive input voltage might cause the transformer to over-heat and burn itself out. If left unattended, a fire could result.

The *secondary voltage* of a power transformer is generally considered the most important specification. It is given in Vac rms, for the stated primary voltage as the input. The secondary voltage rating always assumes the full rated power load is being used—that is, the maximum rated current drain is being used by the load circuit. With a smaller load, the actual secondary voltage will be higher than its rated value.

In selecting a power transformer for a specific application, you must use a somewhat higher secondary voltage than the nominal desired value. For example, if you are building a bridge rectifier power-supply circuit with an intended output voltage of +12 volts, the transformer's secondary will need to put out somewhat more than 12 volts. Why? Because there is a voltage drop across the surge resistor, the filter resistor, and each of the diodes.

The power supply's output voltage will always be somewhat lower than the transformer's secondary voltage, because the power-supply circuit itself needs to use some of the power in order to operate. Manufacturers of transformers take this into account. This is why the standard secondary voltage ratings of off-the-shelf power transformers tend to look so awkward. A transformer rated for 6.3 Vac at its secondary would be used in a 5-volt or 6-volt power supply. The other two most common transformer secondary voltages are 12.6 Vac (for 12-volt power supplies) and 25.2 Vac (for 24-volt power supplies).

In practical design work, the *power rating specification* is almost as important as the secondary voltage. This specification tells us the maximum load the transformer can reliably feed without the risk of burning itself out. The secondary voltage rating assumes the full rated load is being applied to the output of the transformer.

Usually the power rating is given as a wattage, or volts-amperes (VA) value. These two terms are interchangeable, since wattage is equal to the voltage times the current:

$$P = EI$$

To find the maximum acceptable current drain, just divide the power rating by the secondary voltage:

$$I = \frac{P}{E}$$

For example, let's assume we have a power transformer rated for a secondary voltage of 12.6 Vac, and a power rating of 15 watts, or 15 VA. The maximum current drain this transformer can safely handle is about:

$$I = \frac{15}{12.6}$$

$$= 1.2 \text{ amps}$$

The transformer can handle brief load surges of a higher current, as long as this excessive power drain doesn't continue too long.

Some manufacturers and parts dealers give the transformer's power rating directly in current (amperes, or mA ([milliamperes])), instead of in power (watts or VA). For most hobbyist work, this tends to be more convenient. If you need to determine the actual power, you can just multiply the current value by the secondary voltage rating, as suggested earlier:

$$P = EI$$

The regulation factor is probably the least familiar transformer specification for most people working in electronics, whether professionally or on the hobbyist level. It is generally less important than the other specifications we have already considered. In many practical applications, it can safely be ignored.

Earlier, we mentioned that the secondary voltage rating for a power transformer assumes that the full-power load (the power rating) is being applied to the transformer's output (secondary). Reducing the current drain will cause the actual secondary voltage to rise. The regulation factor is a measurement of how much the secondary voltage will increase if the load current is reduced to zero. A typical regulation factor is about 10%. For a 12.6 Vac transformer, with a regulation factor of 10%, with little or no load current being drawn, the actual secondary output voltage could be as high as:

$$V_{out} = V_s + 0.1V_s$$
$$= 12.6 + (0.1 \times 12.6)$$
$$= 12.6 + 1.26$$
$$= 13.86 \text{ Vac}$$

(In this equation V_s stands for the rated secondary voltage of the transformer, and V_{out} is the actual output voltage, with a 0 ([or near 0]) load current.)

If the transformer is wired into the circuit backwards, it will act like a step-up transformer. For example, a standard 12.6 Vac power transformer wired backwards, so a 120 Vac source is applied across its nominal secondary winding (here being used as the primary), will produce an output voltage of 1,143 Vac. Only a small current can be drawn through the step-up transformer, since its power rating remains the same, and power is the product of voltage times current.

Accidentally hooking a power transformer up backwards in a power-supply circuit is liable to destroy every semiconductor and many of the other components in the power-supply circuit and in the load circuit.

For the sake of completeness, we will also mention the *isolation transformer*, which has an equal number of turns in its primary and secondary windings. The output voltage is equal to the input voltage (ignoring some small losses within the transformer itself), but the output circuit will be electrically isolated from the input circuit by the transformer.

Selecting the diodes for use in a power-supply circuit is fairly easy. For the most part, we are just concerned with two key specifications. The first of these is relatively simple, although it could have several different names, which can lead to some confusion:

- PIV (Peak inverse voltage)
- PRV (Peak reverse voltage)
- Maximum reverse voltage
- Maximum reverse-bias voltage

You might encounter other names for the same specification. They all mean the same thing. What is the largest voltage the diode can safely and reliably block when it is reverse-biased? If the applied voltage exceeds this maximum voltage, the diode's pn junction will break down, and start to conduct heavily. The odds are very good that the diode will be permanently damaged or destroyed by this uncontrolled avalanche effect.

In power-supply circuits, the PIV just needs to be higher than the maximum expected input voltage on the primary winding of the transformer. Even the worst possible short circuit is unlikely to feed more voltage through the diode(s) than this. Still, it never hurts to allow a little extra headroom in the PIV specification.

For most power-supply circuits running off of ordinary line current, the diodes should have PIV ratings of at least 200 volts, but 300 volts or 400 volts (or even higher) would offer a little more protection at little or no added expense.

A more critical diode specification is the *maximum* (or peak) *current-handling capability*. All current drawn by the load circuit must pass through the diode(s). For safety and reliability, over-rate the diode on current as much as possible. For example, if the load circuit is expected to draw currents up to 1 ampere, don't use a diode rated for just 1 ampere, or even 1.5 ampere. Use at least a 2-ampere diode, and preferably an even heftier unit.

In some cases, you might need to take the voltage drop across the diode(s) into consideration when designing a power-supply circuit. This is due to the small internal resistance of the pn junction when forward-biased. Usually the voltage drop across a forward-biased diode will be minimal, and can be considered negligible. There is some variation in the exact value, but for most silicon diodes (the type most commonly used in modern electronics), the nominal forward-biased voltage drop is about 0.7 volt-per-diode. If there are multiple diodes in series, the voltage drops will be cumulative. Older type germanium diodes have a lower voltage drop—typically only about 0.3 volt, but they are usually less reliable, and have lower maximum current and PIV ratings than silicon diodes.

In a simple half-wave rectifier circuit, there is only one diode, of course. In a full-wave rectifier circuit there are two, and a bridge rectifier circuit has four. In these multiple diode circuits, all of the diodes should be matched—that is, they should all be of the same type number.

The 1N400x series of diodes is very well suited to power-supply applications. The larger the last digit, the greater the maximum ratings of the device. The 1N4002 is good for low-power circuits, but I'd recommend using at least a 1N4003 or 1N4004 in most power-supply circuits, just to be on the safe side. In some high-power applications, you might need something like a 1N4007. I do not recommend using the 1N4001 in power-supply circuits. The 1N400x series diodes are all rated for 1 ampere maximum continuous current, but they can handle surges of up to 30 amperes. The difference is in the PIV ratings:

1N4001   PIV = 50 volts
1N4002   PIV = 100 volts
1N4003   PIV = 200 volts
1N4004   PIV = 400 volts
1N4005   PIV = 600 volts
1N4006   PIV = 800 volts
1N4007   PIV = 1000 volts

If your power-supply circuit must supply more than about 0.75 ampere, I'd suggest using a heavier-duty diode, such as the 1N540x series, which is rated for 3 amperes continuous, and surges up to 200 amperes:

1N5400   PIV = 50 volts
1N5402   PIV = 200 volts
1N5404   PIV = 400 volts

In multiple-diode circuits, especially bridge rectifiers, the full current load is shared by the diodes to some degree.

For the filter capacitor, as a rule of thumb, the larger the capacitance, the better the filtering. For most rectifier-based power-supply circuits, at least 100 μF should be used for the filter capacitor, and preferably a lot more. 500 μF and 1000 μF capacitors are commonly-used values. Since we are talking about such large capacitances, electrolytic capacitors are usually the only practical choice. Electrolytics are polarized, so they must be installed with care. If installed backwards, they will not work, and will almost certainly be permanently damaged. In addition, there is a fairly good chance that an electrolytic capacitor connected to a reversed polarity for an extended period could explode.

The working voltage of the filter capacitor should be at least twice the nominal output voltage of the power-supply circuit. For example, if you are building a +12-volt power supply, use filter capacitors rated for at least 25 volts. However, do not over-rate the working voltage too much. Some electrolytic capacitors can dry out and age prematurely if they are operated on too low a voltage for an extended period. They can even wear out just sitting on a shelf (applied voltage = 0). Newer devices are less susceptible to such problems, but it is still not a good idea to use a

1,000-volt electrolytic capacitor in a 12-volt circuit. Besides, it would be far more expensive and bulky than it needs to be to do the job properly.

The amount of ripple in a power supply's output signal is directly proportional to the load current, and inversely proportional to the filter capacitor's value. That is, increasing the load current will tend to cause more ripple to appear in the output voltage. Using a larger capacitance value will reduce the ripple for a given amount of load current. This is illustrated in the graph of Fig. 1-25.

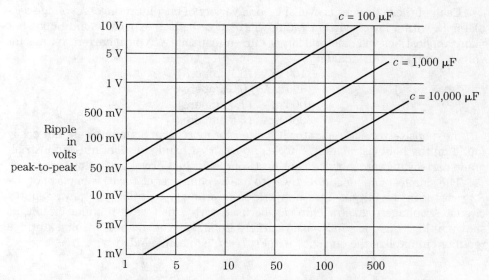

1-25   Using a larger capacitance value will reduce the ripple for a given amount of load current.

Using standard voltage-regulator ICs

Earlier in this chapter, we briefly mentioned voltage-regulator ICs. Now is the time to look at their use a little more closely.

The most popular hobbyist-level voltage-regulator ICs are the 78xx series. There are seven commonly available entries in this series. The "xx" part of the number identifies the regulated output voltage:

7805	5 volts
7806	6 volts
7808	8 volts
7812	12 volts
7815	15 volts
7818	18 volts
7824	24 volts

The output of a 78xx voltage-regulator IC is always positive (with respect to ground or common). There is also a comparable 79xx series that puts out regulated voltages that are negative (with respect to ground, or common):

7905	-5 volts
7906	-6 volts
7908	-8 volts
7912	-12 volts
7915	-15 volts
7918	-18 volts
7924	-24 volts

Each of these devices is available for a variety of current ratings. A voltage-regulator IC with a lower current rating will tend to be less expensive and bulky, but many applications will call for a larger current capability. Typical current ratings for voltage-regulator ICs include:

100 mA	(0.1 ampere)
500 mA	(0.5 ampere)
1000 mA	(1.0 ampere)
3000 mA	(3.0 amperes)

Inexpensive voltage-regulator ICs rated for more than 3 amperes are not easily found on the hobbyist market. Surplus houses that handle discontinued industrial-grade parts would probably be your best source for heftier voltage regulators.

The simplest and most direct way to use a voltage-regulator IC is in place of the R2 resistor in the output filter network of an ordinary rectifier-type power-supply circuit. A voltage-regulator chip can be used with a half-wave rectifier circuit, as shown in Fig. 1-26, or a full-wave rectifier circuit. But for the best results, a bridge rectifier circuit, like the one shown in Fig. 1-27, is recommended.

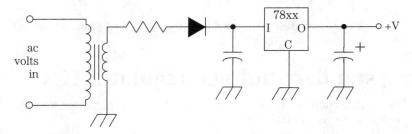

1-26 A voltage-regulator chip can be used with a half-wave rectifier circuit.

Notice that filter capacitors are used on both the input and the output lines of the voltage regulator. The values of these filter capacitors can be much smaller than in an unregulated power-supply circuit. The input capacitor (C1) will usually have a value of about 0.1 µF to 0.5 µF, and should be mounted physically close to the body of the voltage-regulator IC itself. The value of the output capacitor is typically about 10 µF to 100 µF, with values in the lower end of this range being the norm.

To help the voltage regulator do its job most efficiently, a large filter capacitor of the usual type is also commonly used. It is connected in parallel with the voltage regulator's input, as illustrated in Fig. 1-28. As usual, the larger the value of this capacitor, the better. It will usually be mounted at least a couple inches away from the voltage-regulator chip.

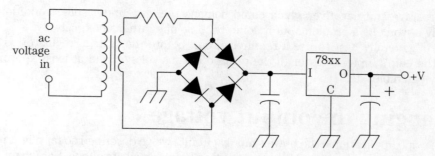

1-27 For the best results, a bridge-rectifier circuit is recommended for use with a voltage-regulator IC.

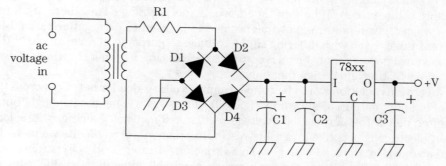

1-28 To help the voltage regulator do its job most efficiently, a large filter capacitor is connected in parallel with the voltage regulator's input.

Some designs omit the rectifiers altogether, and drive the voltage regulator directly from the secondary winding of the power transformer (through a surge resistor), as illustrated in Fig. 1-29—although this probably isn't a good idea in most cases. The voltage-regulator IC will have to work a lot harder than usual in this circuit. At the very least, additional heatsinking should be used. Another disadvantage of this configuration is that the input signal seen by the voltage-regulator IC goes

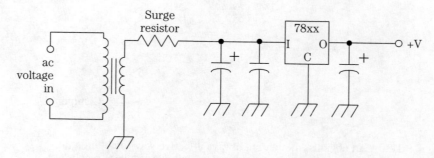

1-29 Some power-supply circuits omit the rectifiers altogether, and drive the voltage regulator directly from the secondary winding of the power transformer through a surge resistor.

both positive and negative, which could damage the chip under some conditions. Usually it won't be a problem, but it could become one rather unexpectedly.

It is generally best to use a rectifier circuit before the voltage regulator so at least the chip's input is closer to the desired output voltage, and at least approximately resembles dc.

Changing the output voltage

In some applications, the power-supply circuits we've described so far might not be satisfactory because they do not put out the desired level of voltage. For example, you might want to build a regulated power supply that will drive a 9-volt device, but there is no 7809 voltage-regulator IC—you have to use either a 7808 (8 volts) or 7812 (12 volts), and neither could be close enough. Some applications might require a supply voltage that can be fine-tuned over a specific range. And there are a few cases when a given power supply might need to be used to drive different loads at different times, each with differing supply voltage requirements. An example of this is a universal power supply on an electronics technician's workbench, which is used for various testing and design functions.

An obvious solution would be to add a simple voltage-divider network across the power supply's output, as shown in Fig. 1-30. Often this will be good enough, but it can often defeat the purpose of a voltage regulator somewhat. Remember, the load circuit electrically acts like a variable-resistance element, changing its value as the current drawn by the load changes. This variable resistance load is in parallel with the lower half of the voltage divider network, inevitably affecting its value, causing the supplied voltage to fluctuate with changes in the current drawn by the load. Such a crude solution therefore usually won't be too helpful in most practical applications.

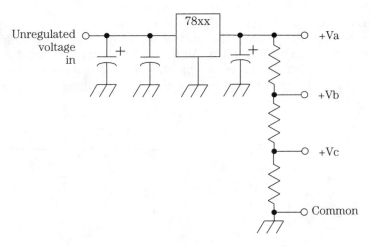

1-30   An obvious, though not always effective way to obtain a different output voltage from a fixed-voltage regulator chip is to add a simple voltage-divider network across the power supply's output.

A simple 78xx series voltage-regulator IC has just three leads, and is so simple, it would seem you'd be pretty much stuck with its designed output voltage. After all, there is no "voltage adjust" pin or anything like that. These chips were designed for fixed voltages, but they can be tricked into operating at voltages somewhat different from their original design value.

The secret is in where you put the true ground in the circuit. The output voltage of a three-pin voltage-regulator IC is referenced to its common pin. Ordinarily, this pin is connected to true ground potential, so the output voltage from a 7805, for example, is five volts above ground—that is, just plain +5 volts.

Most voltage-regulator ICs draw only a small quiescent current (typically just a few milliamperes) that flows to ground from the common pin. By applying a bias voltage to the common pin, we can "float" it above ground, and fool the voltage regulator into putting out a higher-than-normal output voltage.

The simplest way to do this is to add a variable resistance (normally a potentiometer of some sort) between the common pin and true circuit ground, as illustrated in Fig. 1-31. If you don't need a true variable output voltage, just an "oddball" value, you could substitute a fixed resistor with an appropriate value. The best and most reliable way to find the correct resistance is to temporarily hook up a potentiometer, as shown here, and adjust it carefully for the desired output voltage. Then, without moving the potentiometer's shaft, carefully measure its resistance value. You will probably need to use a precision resistor to come close enough to the desired value.

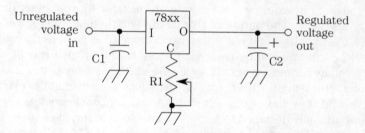

1-31   The easiest way to "fool" a voltage regulator into putting out a higher than normal output voltage is to add a variable resistance between the common pin and true circuit ground.

This common-to-ground resistance should not be very large. I would not recommend using much more than a 1-kΩ (1,000 ohms) potentiometer. You can "fool" a 78xx just so much before the trick ceases to work reliably, and the IC could possibly be damaged.

This circuit is simple and inexpensive, but the regulation of the output voltage can be adversely affected. The output voltage will change with any shifts in the quiescent current, which can occur for a variety of different reasons.

An improved variable output voltage-regulator circuit is shown in Fig. 1-32. Here we have added a second, feedback resistor (R2) from the chip's output to its common pin. A value of about 1 kΩ (1,000 ohms) should be used in most cases. You might want to decrease the maximum value for potentiometer R1 just to be safe. Try using a 500-Ω potentiometer in this circuit. Yes, you will get a smaller range of out-

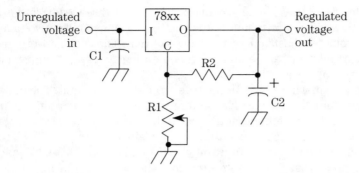

1-32 This is an improved variable-output voltage regulator circuit.

put voltages with a smaller potentiometer, but this is a reasonable trade-off for significantly improved regulation of the output voltage.

For even greater stability when adapting a standard voltage-regulator IC for a fixed "oddball" output voltage, you can connect a zener diode from the common pin to circuit ground as shown in Fig. 1-33. An output to a common feedback resistor should always be used in this case, probably with a somewhat higher value, say, around 3.3 kΩ (3,300 ohms) to 4.7 kΩ (4,700 ohms). The circuit's output voltage will be equal to the sum of the voltage-regulator IC's nominal output voltage, and the zener diode's avalanche voltage. For example, let's say we are using a 4.2-volt zener diode with a 7815 voltage regulator. The output of this circuit will then be a fairly well-regulated 19.2 volts (4.2 + 15 volts). The regulation won't be quite as good as the voltage regulator by itself, but it will still be reasonably close to the original specifications, and should be adequate for most practical applications. It's certainly better than nothing.

Semiconductor manufacturers quickly recognized the usefulness of such circuit design tricks, and they soon came out with special adjustable voltage-regulator ICs that were intentionally designed to accommodate such external changes to the output voltage. One of the simplest devices of this type is the LM317, which has three terminals just like a 78xx voltage regulator, except instead of INPUT—COMMON—OUTPUT, the pin functions of the LM317 are INPUT—ADJUST—OUTPUT.

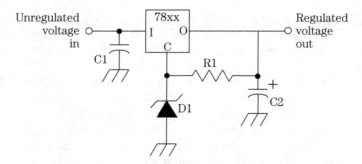

1-33 Better variable voltage regulation can be achieved by adding a zener diode.

The basic LM317 variable-output voltage-regulator circuit is shown in Fig. 1-34. It can be operated reliably over a much wider range and with better regulation than a 78xx chip or other fixed voltage-regulator device. The LM317 is designed to accept input voltages from 4 to 40 volts, and can put out regulated voltages ranging from 1.25 to 37 volts. The current rating for this IC is 1.5 amperes.

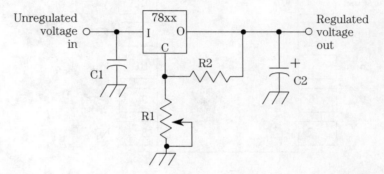

1-34   This is the basic LM317 variable-output voltage regulator circuit.

A similar device is the LM338, which can handle currents up to 5 amperes. The input voltage range is the same as for the LM317, but the maximum regulated output voltage is just 32 volts for the LM338. The lower end of the output voltage range is still 1.25 volts.

A somewhat more sophisticated and versatile variable-voltage-regulator IC is the 723. This chip is housed in a standard 14-pin DIP package, but three of the pins are not internally connected to anything. The pin-out diagram for the 723 is shown in Fig. 1-35.

This chip will accept unregulated input voltages of up to 40 volts, and its regulated output voltage can be anything from 2 to 37 volts. Normally, the maximum output current for the 723 is just 150 mA (0.15 ampere), but it is a fairly simple matter to add some external power transistors as current amplifiers to supply currents of up to 10 amperes.

For a variety of technical reasons, slightly different circuitry should be used for low-output voltages (2 to 7 volts) than for higher-output voltages (7 to 37 volts). The low-voltage version is shown in Fig. 1-36.

The reference voltage (V_{ref}) is fed into pin 6, through resistors R1 and R2. This reference voltage will normally be between 6.8 volts and 7.5 volts. The formula for this circuit's output voltage is:

$$V_{out} = \frac{(V_{ref} \times R_2)}{(R_1 + R_2)}$$

The value of resistor R3 should be equal to the parallel combination of R1 and R2. That is:

$$R_3 = \frac{(R_1 \times R_2)}{(R_1 + R_1)}$$

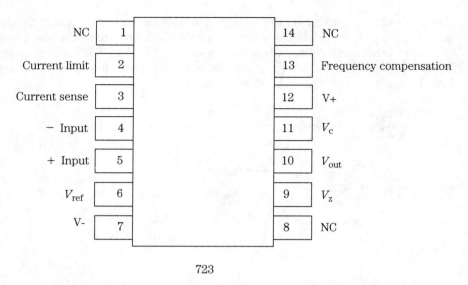

1-35 A somewhat more sophisticated and versatile variable voltage-regulator IC is the 723.

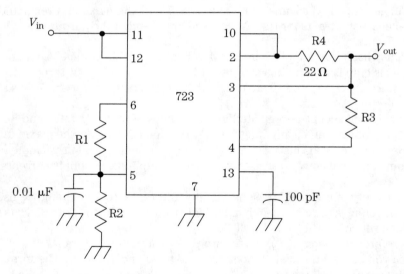

1-36 This 723 circuit can be used to generate low output voltages (from 2 to 7 volts).

For best results, precision resistors are recommended, but for general purpose applications, standard 5% resistors should be close enough. Standard resistor values for some typical output voltages are summarized in Table 1-3. Remember, these resistor values are rounded off to the nearest standard values, so the output voltages will not be exact.

Table 1-3. Standard resistor values for some typical output voltages for the low-voltage 723 power-supply circuit of Fig. 1-35

Output voltage	R_1	R_2	R_3
3.0	4.2 kΩ	3.3 kΩ *	1.8 kΩ
3.5	3.6 kΩ	3.6 kΩ	1.8 kΩ
5.0	2.2 kΩ	4.7 kΩ	1.5 kΩ
6.0	1.2 kΩ	6.2 kΩ	1 kΩ

*3.0 kΩ is better, if available

The 7- to 37-volt version of the 723 variable voltage-regulator circuit is shown in Fig. 1-37. Basically, we've just moved the resistors around some. The reference voltage (V_{ref}) is still on pin 6, and resistor R3 is still equal to the parallel combination of R1 and R2:

$$R_3 = \frac{(R_1 \times R_2)}{(R_1 + R_2)}$$

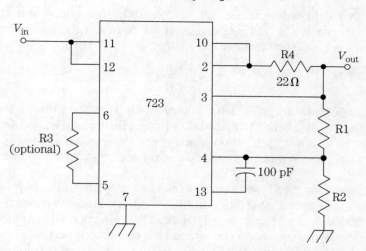

1-37 This 723 circuit can be used to generate higher output voltages (from 7 to 37 volts).

In some applications, resistor R3 is optional, but including it gives the circuit better temperature stability. Resistors are so inexpensive, I see little reason to want to omit this component.

The output voltage formula for this version of the 723 is a little different:

$$V_{out} = \frac{V_{ref} \times (R_1 + R_2)}{R_2}$$

Once again, for the best possible results, precision resistors are recommended, but for general-purpose applications, standard 5% resistors should be close enough. Standard resistor values for some typical output voltages in this circuit are summarized in Table 1-4. Remember, these resistor values are rounded off to the nearest standard values, so the output voltages will not be exact.

Table 1-4. Standard resistor values for some typical output voltages for the high-voltage 723 power-supply circuit of Fig. 1-36

Output voltage	R_1	R_2	R_3
9	1.8 kΩ	7.2 kΩ *	470 Ω
12	4.7 kΩ	7.2 kΩ	2.7 kΩ
15	7.5 kΩ	7.2 kΩ	3.6 kΩ
29	22 kΩ	7.2 kΩ	5.6 kΩ

*6.8 kΩ or 7.5 kΩ could be used if 7.2 kΩ is not available

Current regulation

Ordinarily, the load determines how much current it will draw from the power supply in response to the voltages and resistances within the load. The current drawn by the load is determined by Ohm's law

$$\left(I = \frac{E}{R} \right).$$

In some specialized applications, however, we want the current level to be consistent and independent of the load. In such applications, we need a fixed current source, which always puts out a specific, pre-determined amount of current. This type of circuit is sometimes called a *constant current source*, or a *current limiter*.

Roughly speaking, this type of circuit could be considered a *current regulator*, analogous to a voltage regulator. The output current in this case is regulated, or held to a specific value, and is not permitted to fluctuate during normal operation. A simple, but practical fixed current source circuit built around one section of an LM3900 quad Norton amplifier IC is illustrated in Fig. 1-38. A typical parts list for this circuit is given in Table 1-5.

Because the output current from this type of circuit is related to the voltage, the actual power supply should be well-regulated, with the voltage regulation circuitry coming before the current source circuit.

Throughout the following discussion, we will assume that a +15-volt voltage regulator is being used to drive the fixed current source circuit. Using the component values from the suggested parts list of Table 1-5, the output current from this circuit will be fixed at 1 mA (0.001 ampere).

The load (RL) being driven by the fixed current must be connected between the collector of output transistor Q1 and ground. The current source will hold its con-

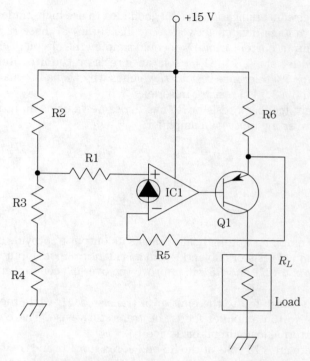

1-38 A LM3900 Norton amplifier can be used as the
heart of a simple, but practical fixed-current
source circuit.

**Table 1-5. Suggested parts list for for the
fixed-current source circuit of Fig. 1-37**

IC1	LM3900 quad Norton amplifier (one section only)
Q1	pnp transistor (2N3906, or similar)
R1, R5	1-Ω, 5%, ¼-W resistor
R2, R6	1-kΩ, 5%, ¼-W resistor
R3	10-kΩ, 5%, ¼-W resistor
R4	3.9-kΩ, 5%, ¼-W resistor

stant value as long as the load impedance is no greater than about 14 kΩ (14,000
ohms). If the load impedance is higher than this, the actual current value will drop.
But the current will never exceed its rated, constant value, regardless of the load resistance. The load impedance might drop all the way down to 0 ohms (a dead short)
without affecting the current source.

The input to the Norton amplifier (IC1) is derived from a simple voltage divider
network made up of resistors R2, R3, and R4. The values of these resistors are selected to present a 14-volt input to the Norton amplifier's non-inverting input,
through input resistor R1.

For high-precision applications, it is a good idea to use high-grade, 1% tolerance resistors in this voltage-divider network. Any inaccuracy in these resistance values will affect the output current value. You could simplify the circuit slightly by replacing R3 and R4 with a single 14-kΩ 1% tolerance resistor. Unfortunately, this is not a standard value for 5% tolerance resistors, which is why we had to make it up from a 10-kΩ resistor and a 3.9-kΩ resistor in series.

In this circuit, feedback resistor R5 has the same value as input resistor R1, so we have a non-inverting unity gain amplifier:

$$G = \frac{R_5}{R_1}$$

$$= \frac{1,000,000}{1,000,000}$$

$$= 1$$

The Norton amplifier automatically adjusts its output to provide an output voltage at the junction of resistors R5 and R6, that is identical to its input voltage. Again, using 1% resistors for R1 and R5 will improve the overall accuracy and precision of the circuit.

This voltage is also fed to the emitter of transistor Q1, while the direct output signal from the Norton amplifier feeds the transistor's base. The collector is connected to the external load circuit or device.

Because there are +14 volts at the R5 end of resistor R6, and the full supply voltage (+15 volts) at the other end of this resistor, it necessarily follows that the voltage drop across this component must always be 1 volt. Knowing the value of resistor R6 (from the parts list), we can now use Ohm's law to find the current flowing through it. For greatest accuracy, a 1%-tolerance resistor is recommended for R6 too.

According to the parts list, resistor R6 has a value of 1 kΩ (1,000 ohms). The current flowing through this component therefore works out to:

$$I = \frac{E}{R}$$

$$= \frac{1}{1000}$$

$$= 0.001 \text{ ampere}$$

$$= 1 \text{ mA}$$

This 1-mA current is derived from the emitter of transistor Q1. You should recall from your basic electronics theory that a transistor's emitter and collector currents are virtually identical, so the output current to the load (RL) is also about 1 mA, and is held at that constant value, regardless of any small-to-moderate fluctuations in the load impedance of RL.

The fixed-output current from this circuit can be increased by decreasing the value of resistor R6. Reducing this resistance by half doubles the output current. It is a fairly simple matter to use Ohm's law to calculate the necessary value of resistor R6. Assuming that none of the other values throughout the fixed current source circuit are changed, the formula is:

$$R = \frac{E}{I_d}$$

$$= \frac{1}{I_d}$$

The voltage drop across R6 should always be 1 volt, assuming all other resistor values (and the supply voltage) remain the same. I_d is the desired output current value in amperes (not mA).

Do not try to make the output current value for this circuit too large, or the transistor, the IC, or both, could be damaged. Check the manufacturer's specification sheets for the particular semiconductor components you are using in your circuit to determine the maximum safe output current value. Be sure to leave some headroom—don't try to use a current value right at the component's absolute maximum limit.

Transistor Q1 can be almost any standard pnp-type unit. The 2N3904 recommended in the parts list is a good, general-purpose choice. Just make sure that the transistor you select for use in this circuit can safely and reliably handle the desired output current. As a general rule of thumb, over-rate the transistor's current-handling capability by at least 20% to 30%. That is, if the manufacturer says the absolute maximum current for your transistor is 2.5 amperes, treat it as if the practical maximum current value is somewhere between 1.75 amperes and 2.0 amperes.

Project #1—Multiple-output, dual-polarity power supply

I have already given all the significant technical details for the two power supply projects of this chapter. In both projects, we are simply putting together several circuits that have already been described.

In certain applications, multiple-supply voltages, perhaps both positive and negative, might be needed by different sections of the load circuitry. Our first project is designed to simultaneously put out three positive voltages and three negative voltages, each individually regulated.

The schematic diagram for this multiple-output dual-polarity power supply project is shown in Fig. 1-39. The suggested parts list for this project is given in Table 1-6.

Notice that a center-tapped power transformer is required in this project to permit both positive-(above ground) and negative-(below ground) output voltages. The transformer's center-tap is grounded. The full secondary voltage of the power transformer must be a little more than twice the largest desired output voltage, which we are assuming to be 15 volts. Fifteen volts above ground (+15 volts) plus fifteen volts below ground (–15 volts) equals a total of 30 volts peak-to-peak. The nearest standard transformer voltage is 36 volts, which is just about perfect for our purposes here.

For the time being, let's just consider the positive side of the circuit. The unregulated positive voltage is tapped off the bridge at the junction between D2 and D4. C1 is a large filter capacitor. It's exact value is not critical, but the larger it is, the better job of filtering it will do. The nominal dc voltage at this point is approximately 17 volts—one half the voltage of the secondary winding (referenced to ground) less the

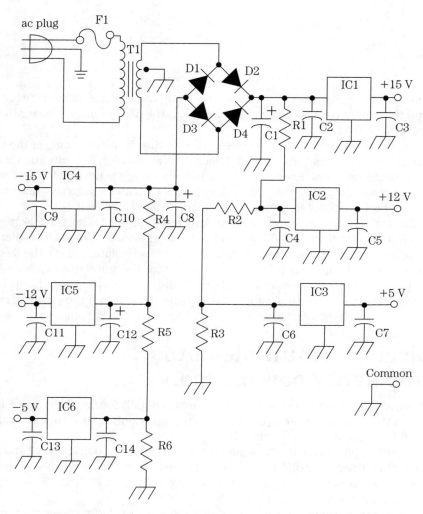

1-39 Project # 1—Multiple-output, dual-polarity power supply.

Table 1-6. Suggest parts list for Project #1—
Multiple-output, dual-polarity power-supply of Fig. 1-39

IC1	7815 +15-V, 500-mA voltage-regulator IC
IC2	7812 +12-V, 500-mA voltage-regulator IC
IC3	7805 +5-V, 500-mA voltage-regulator IC
IC4	7915 –15-V, 500-mA voltage-regulator IC
IC5	7912 –12-V, 500-mA voltage-regulator IC
IC6	7905 –5-V, 500-mA voltage-regulator IC
D1, D2, D3, D4	1N4003-diode
T1	Power transformer—secondary 36 Vac, center-tapped

F1	4-A fuse and holder
R1, R4	2.7-kΩ, 5%, ¼-W resistor
R2, R5	8.2-kΩ, 5%, ¼-W resistor
R3, R6	6.8-kΩ, 5%, ¼-W resistor
C1, C2	2,500-μF, 25-V electrolytic capacitor
C2, C4, C6, C9, C11, C13	0.22-μF capacitor
C3, C5, C7, C10, C12, C14	10-μF, 20-V electrolytic capacitor

normal voltage drop across the active bridge diodes. Let's mentally disconnect resistor R1 from the circuit for now. This means there is no current path to the circuitry surrounding IC2 and IC3—they are not part of the circuit until R1 is replaced.

This leaves us with just a simple, standard-voltage circuit, built around IC1. This IC is a 7815, so it's regulated-output voltage is +15 volts. Capacitors C2 and C3 are just the standard filter capacitors, almost always used with 78xx series voltage regulators.

So far, we have nothing unusual. Now, let's mentally reconnect resistor R1 to the circuit. This resistor, along with R2 and R3, forms a simple resistive voltage-divider network. Using the component values suggested in the parts list, IC2 (a 7812) sees an unregulated-input voltage of a little less than +14.5 volts, and puts out a regulated +12 volts. Similarly, IC3 (a 7805) sees an unregulated input voltage of a little more than +6.5 volts, and puts out a regulated +5 volts.

Because the resistive-voltage divider network comes before the voltage regulators, there are no loading effects, and fluctuations in the load circuit make no difference.

Strictly speaking, these resistors aren't absolutely necessary. Even a 7805 can take an unregulated input voltage up to about 30 volts, but there seems little reason to make the voltage-regulator chip work that hard. The three resistors are likely to be considerably less expensive and bulky than the additional heatsinking that might be required without them.

If you are using voltage regulators with larger output currents than what is recommended in the parts list, you might need to use resistors with higher wattage ratings. If the resistor can't handle the current drawn through it, it could change value, which usually won't be too much of a problem in this particular application. More seriously, the resistor could burn itself out totally, and act essentially like an open circuit. Any later voltage-regulator stages won't receive any input voltage, so that output will be dead.

The negative-output voltage section of the circuit works in just the same way, except 79xx voltage-regulators are used in place of the positive 78xx devices of IC1, IC2, and IC3. The unregulated negative voltage is tapped off between bridge diodes D1 and D3.

The input fuse (F1) should be selected to handle a little more than the sum of the maximum output currents for each voltage regulator. Since the parts list recommends 500-mA voltage regulators, and there are six of them, this is a total acceptable current of 3 amperes. A 3.5-ampere to 4-ampere fuse will offer sufficient headroom, but still should blow before any damage is done in case of a short circuit in the power supply or in the load circuit. For greater protection, you might want to add additional fuses in each of the output lines. Using the recommended voltage-regulators, each output fuse should be rated for ½ ampere (500 mA). Use automotive fuses here, because they are designed for lower dc voltages. A regular 120-volt fuse might not blow in time, even if the rated current value is exceeded.

Project #2—Variable-output, current-limited power supply

Our next power supply project permits manual adjustment of the output voltage. In addition, the output current is limited, and the user can manually adjust the maximum output current as well. The schematic diagram for this project is shown in Fig. 1-40, with a suitable parts list appearing in Table 1-7.

Once again, we are just putting together a couple of the basic circuits discussed earlier in this chapter. This project is designed around an LM317 adjustable voltage-regulator IC (IC1). We are using just the basic LM317 variable output voltage circuit here, except for the addition of diode D5 and capacitor C3, which improves the stability and reduces the output ripple. This circuit can offer a ripple rejection figure of up to 80 dB, which is excellent for most practical purposes.

Another addition to the basic LM317 circuit here is meter M1. This is just a small dc voltmeter, to permit the user to know what output voltage the power supply is currently set for. The output voltage is adjusted via potentiometer R3, which should be a front panel control, mounted as close as possible to M1 for convenience.

The output of the LM317 voltage-regulator circuit is then fed through the same current-limiter/fixed current-source circuit we discussed in an earlier section. A milliammeter (M2) is added in series with the current determining resistance, which in this circuit is comprised of the series combination of R8 and R9. R9 is another front panel mounted potentiometer, near M2. Resistor R8 is included to prevent the possibility of setting R9 too close to zero.

Because of the placement of the milliammeter (M2) in this circuit, the actual output current being drawn by the load circuit is not indicated. The actual load current might be lower than the reading on the meter, but it won't be permitted to exceed it. I believe this is more useful in practical applications. The user can adjust R9 for the desired current limit, without worrying about what the load impedance is at the moment. There is also less possibility of loading problems in this arrangement.

In use, adjust the desired output voltage first (via R3), then select the desired current limit (via R9). Do not reverse this sequence. The current limit is dependent on this circuit's input voltage. Changing the voltage setting, without moving the shaft of R9, will result in a different current-limit setting.

Because of the inherent inaccuracies of this voltage dependence, there wouldn't be much point in using high-precision 1% tolerance resistors for R4 through R8. Any inaccuracies in these resistances can be compensated for by adjusting potentiometer R9 until the desired current value is read on the milliammeter (M2).

The three extra sections of the quadNorton amplifier chip (IC2) can be left disconnected, or they can be used independently (except for the supply voltage) in other circuitry as part of a larger system.

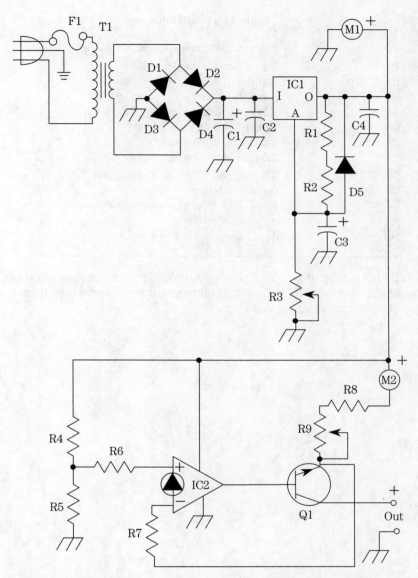

1-40 Project #2—Variable-output/current-limited power supply.

**Table 1-7. Suggested parts list for Project #2—
Variable-output, current-limited power supply of Fig. 1-39**

IC1	LM317K adjustable voltage-regulator IC
IC2	LM3900 quad Norton amplifier (one section only)
Q1	pnp transistor (2N3904, or similar)
D1, D2, D3, D4, D5	1N4003 diode

Table 1-7. Continued

T1	Power transformer—secondary 40 Vac
F1	3-amp fuse and holder
R1, R2, R8	100-Ω 5%, ¼-W resistor
R3	5-Ω potentiometer
R4	1-kΩ, 5%, ¼-W resistor
R5	10-kΩ, 5%, ¼-W resistor
R6, R7	1-MΩ, 5%, ¼-W resistor
R9	2.5-kΩ potentiometer
C1	2,200-µF, 50 V electrolytic capacitor
C2	0.1-µF capacitor
C3, C4	10-µF, 50 V electrolytic capacitor
M1	dc voltmeter 0–50 V
M2	dc milliammeter 0–1 amp

This power supply project is designed to provide positive-regulated output voltages ranging from +2 to +37 volts. Major changes in the design would be needed to accommodate negative-output voltages.

2
CHAPTER

Audio oscillators

In practice, the terms *oscillator* and *signal generator* are basically interchangeable. Both generate ac waveforms. But to be technically accurate, there is a difference, which is based on the waveform generated. Strictly speaking, if the circuit puts out only sine waves, it is an oscillator. If it puts out any other waveform in addition to, or instead of a pure sine wave, then it is more properly referred to as a signal generator.

In this chapter we will focus our attention on true sine-wave oscillators, which operate primarily in the audio frequency range—nominally from about 20 Hz up to approximately 20 kHz (20,000 Hz).

The *frequency* of any repeating ac waveform is the number of complete cycles that pass a given point per second. In fact, frequency is measured in *cycles per second*, usually abbreviated as *cps*. More recent technical literature usually gives frequency values in *Hertz*, abbreviated as *Hz*. The newer name is in honor of an important scientist from the early days of electronics. The two terms Hertz and cycles per second mean exactly the same thing, and can always be considered fully interchangeable.

A *sine wave* is the simplest possible ac waveform. A pure, undistorted sine wave consists of a single frequency component, known as the *fundamental frequency*. There is no harmonic content or any overtones at all. We'll get into the concepts of harmonics and overtones in later chapters.

Theoretically, any ac waveform can be created by an appropriate combination of sine waves, at various frequencies and amplitudes. The process of combining sine waves to create more complex waveforms is called *additive synthesis*.

The sine wave gets its name from its appearance when it is displayed on an oscilloscope, as illustrated in Fig. 2-1. This waveshape resembles a graph of the sine function in trigonometry.

In practice, it is very difficult to generate a truly pure sine wave. Usually there will be some distortion, which means the addition of extra frequency components. However, a reasonably good circuit design will minimize the amplitude of these dis-

2-1   The simplest ac waveform is the sine wave.

tortion frequency components. It is not difficult to generate a sine wave that is pure enough for most practical applications.

High-quality sine-wave signals are often used for electronics testing because the purity of the source waveform makes it easy to spot many effects of distortion in the tested circuit. For example, Fig. 2-2 shows a typical test set-up for an audio amplifier. Ideally, a dual-trace oscilloscope should be used, so the input and output signals can be directly compared instant for instant. The output waveform should be identical to the input waveform, except for the increased amplitude (boosting the signal amplitude is the purpose of an amplifier, of course). Any differences in waveshape are due to distortion in the amplifier's circuitry.

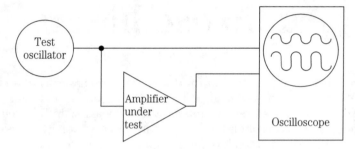

2-2   Sine waves are often used in testing applications, like checking the distortion of an audio amplifier.

LC parallel-resonant tanks

Most sine-wave oscillator circuits are built around parallel-resonant LC circuits (or their equivalent). This is nothing more than an inductor (or coil) and a capacitor connected in parallel. The fundamental operation of a basic LC oscillator tank is illustrated in Fig. 2-3. This is not a practical circuit, just a theoretical illustration of the basic principles involved.

Initially, with the switch open, as shown in Fig. 2-3A, nothing at all happens. No current can flow through the input circuit, so there is no power within the secondary tank.

When the switch is closed, as in Fig. 2-3B, the voltage through coil L1 (the primary coil of a transformer) rapidly increases from zero up to the source voltage. This means that the current through this coil must also be increasing proportionately. According to the principle of electromagnetic induction, a change of current flowing through a coil generates a *magnetic field*. This magnetic field induces a similar voltage in coil L2 (the secondary winding of the transformer), which is stored by capacitor C.

Because we are switching in only a dc voltage, the current through L1 very quickly reaches a stable point, then stops increasing, so the magnetic field generated around this coil collapses. This means that no further voltage is induced into L2, and

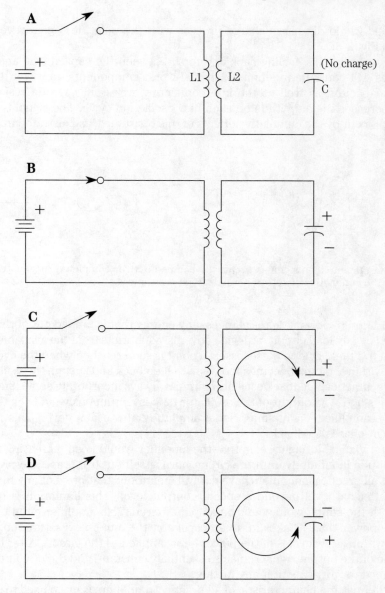

2-3 Most simple sine-wave oscillator circuits are built around a parallel-resonant LC tank, (or its electrical equivalent).

C can begin to discharge through the coil in the opposite direction from the original voltage.

This discharge voltage now causes L2 to induce a voltage back into L1, which in turn, induces the voltage back into L2, recharging C, but now with the opposite polarity as before, as illustrated in Fig. 2-3C. Once the induced voltage in the coils collapses, this whole discharge/charge process with the polarities again reversed, as shown in Fig. 2-3D.

Theoretically, this cycling back and forth between the capacitance and the inductances will continue indefinitely. In real-world components, however, the coils and the capacitor, as well as the interconnecting wires, all have dc resistance—which decreases the amplitude on each new oscillation cycle. The signal is eventually damped out by the cumulative effects of this resistance, as illustrated in Fig. 2-4.

2-4   An unamplified signal is eventually damped out (or completely attenuated) by the cumulative effects of normal circuit resistances.

In addition to normal dc resistances, any energy that is tapped out of the circuit to be used by any load circuit or device obviously subtracts from the available energy within the LC tank. Due to these losses, a point is soon reached when the overall signal level within the tank becomes too weak to feed back and sustain the oscillations. Therefore, practical feedback oscillator circuits always incorporate some kind of amplification stage. The output of the oscillator tank is continuously fed back to the input of the amplifier, maintaining signal amplitude at a usable level. This is called positive (in-phase) feedback.

At first glance, it might seem that the amplifier would keep increasing the output amplitude indefinitely until the circuit burns itself out from the excessive signal level. But all practical amplifiers have natural limitations that prevent any further increases in signal level beyond a specific output level. This limitation is normally linked with the supply voltage of the amplifier circuit. No amplifier circuit can put out more power than it takes in. In an oscillator, the amplifier's saturation point is reached within a few cycles of the power being applied to the circuit. After that, the amplitude of the output signal remains essentially constant, and the circuit designer doesn't need to worry about it any further.

Another natural characteristic of all practical amplifiers is often used to start oscillations in the first place. All amplifiers inevitably generate some internal noise and produce a tiny output signal—even if the input is perfectly grounded (absolutely no input signal at all). This noise can be fed back through the amplifier's input until its amplitude has been increased enough to start oscillations within the LC tank.

The frequency of the signal generated by this type of oscillator is determined by the resonant frequency of the specific coil-capacitor combination. Any combination of an inductance and a capacitance has one (and only one) specific resonant frequency, at which the combination will naturally respond with maximum energy transfer efficiency. The standard LC resonant frequency formula is:

$$F = \frac{1}{(2\,\pi\,\sqrt{LC}\,)}$$

where F is the frequency in hertz, L is the inductance in henries, and C is the capacitance in farads. As in all electronics equations, always be careful to use the correct units for each value, or you will not get the correct results.

Since pi (π) is a universal constant that always has a value of approximately 3.14, we can re-write the formula as:

$$F = \frac{1}{(6.28\,\sqrt{LC}\,)}$$

As an example, let's say we have a 150-mH (0.15 H) coil and a 50-μF (0.00005 farad) capacitor. The resonant frequency of this combination works out to:

$$F = \frac{1}{(6.28\,\sqrt{(0.15 \times 0.00005)}\,)}$$

$$= \frac{1}{(6.28\,\sqrt{0.0000075}\,)}$$

$$= \frac{1}{(6.28 \times 0.002739)}$$

$$= \frac{1}{0.0172}$$

$$= 58 \text{ Hz}$$

Decreasing either the inductance or the capacitance (or both) increases the resonant frequency, and vice versa.

Many standard sine-wave oscillator circuits are built upon these basic principles. Most, though not all, are named after their original designers. We will briefly look at just a few of the more important circuits of this type before we get to our actual audio-frequency oscillator project.

The Hartley oscillator

One of the most common simple oscillator circuits is the *Hartley oscillator*, shown in basic form in Fig. 2-5. This circuit is often referred to as a *split-inductance oscillator* because coil L is center-tapped, making it a sort of autotransformer. In effect, coil L acts like two separate coils in very close physical proximity. A current through coil section AB induces a proportional signal into coil section BC, and vice versa.

When power is first applied to this circuit, resistor R2 places a small negative voltage on the base of the transistor, allowing it to conduct. Internal noise builds up within the transistor amplification stage. When this signal reaches a usable amplitude, current from the transistor's collector passes through resistors R4, R2, and R1. This rising current finally reaches coil section AB, inducing a comparable voltage into coil section BC. The induced voltage is stored by capacitor C1.

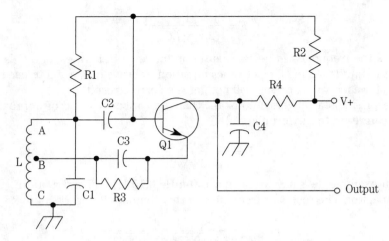

2-5   One of the most common and simple oscillator circuits is the Hartley oscillator.

Capacitor C2 is selected to have a very low impedance at the oscillating frequency. In effect, the transistor's base is more or less directly connected to C1. The base voltage provided by R2 is quite low, so it can be reasonably ignored once oscillations have been started. That connection is needed only to initiate the oscillation process when the circuit is first powered up.

As capacitor C2 charges up, it increases the bias on the transistor. This, in turn, increases the current through coil section AB, and the induced voltage through coil section BC, and the charge on both C1 and C2 is also increased.

Eventually (actually within a few fractions of a second), the voltage stored across C1 equals the $R_1 \backslash C_2$ voltage, but with the opposite polarity. These two voltages now completely cancel each other out. At this point the transistor is saturated, and its output current stops rising, so the magnetic field around coil section AB collapses. No further voltage is induced into coil section BC.

Capacitor C1 now starts to discharge through coil section BC, allowing capacitor C2 to discharge through resistor R1, which cuts off the transistor until the next cycle begins.

Naturally, it takes some finite time for C1 to discharge through coil section BC. Therefore, as the current through the coil is increased, it builds up an electromagnetic field. Once the capacitor is sufficiently discharged, it stops supplying current to the coil—but the coil tends to oppose the change in current flow. It continues to conduct for a brief time, charging the capacitor in the opposite direction. The new voltage across C1 turns on the transistor, and the entire process is repeated.

In some practical applications, the low impedance of the transistor may load the tank circuit excessively, increasing power loss, and possibly decreasing the stability of the circuit. This problem can be overcome by using a FET or some other high-impedance active device in place of the bipolar transistor.

The Colpitts oscillator

A close relative of the Hartley oscillator is the *Colpitts oscillator*. It is illustrated in basic form in Fig. 2-6. Where the Hartley oscillator is based on a split-inductance, the Colpitts employs a split-capacitance. Otherwise, they are quite similar.

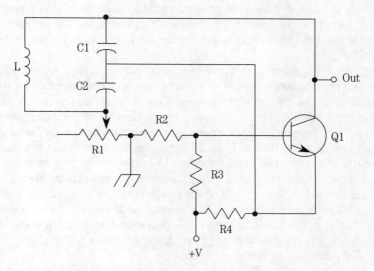

2-6   The Hartley oscillator is closely related to the Colpitts oscillator.

The two capacitors (C1 and C2) are connected in series, and function as if they were a single capacitor—as far as the LC resonant tank is concerned. But there is a center tap (the connection point between the two individual capacitors), which provides a feedback loop path to the transistor's emitter through the four resistors.

If the two capacitors are of equal value, the total effective capacitance within the LC tank (which determines the resonant frequency) is equal to one half the value of either capacitor separately.

If the two capacitors in the tank have different values, the total effective capacitance can be calculated with the standard formula for two capacitances in series:

$$C_t = \frac{(C_1 \times C_2)}{(C_1 + C_2)}$$

In most practical Colpitts oscillator circuits, the two capacitors generally are not equal. This is because the strength of the feedback signal is dependent on the ratio of these two capacitances. By changing both of these capacitor values in inverse fashion (increasing one, while decreasing the other by a like amount), the feedback level can be varied, while the circuit's resonant frequency is held constant.

This brings up one of the chief limitations of the Colpitts oscillator. It is not always convenient to change the oscillation frequency while it is in operation. When the frequency is changed, you obviously don't want the amplitude of the feedback signal to vary, or the amplitude of the oscillator's output signal will not be constant. In some cases, oscillations may not be sustained in the circuit. This is not really an insurmountable problem, just an inconvenience.

A number of solutions have been found by various circuit designers over the years. Basically, both capacitances must be changed simultaneously. The most obvious approach is to find a reverse-ganged variable capacitor. Unfortunately, such components are quite rare, and hard to locate (especially on the hobbyist market), although you could get lucky with a surplus dealer who carries industrial equipment and parts.

It might seem logical that you could just make the coil adjustable instead of the capacitors. After all, changing either the inductance or the capacitance in the tank permits control of the resonant frequency. This is true enough on the theoretical level, but in most applications, it is just as impractical. Without getting into unnecessary detail here, a general rule of thumb is that it is almost always preferable to use an adjustable capacitor rather than an adjustable inductor whenever possible.

A frequently-used solution in practical Colpitts oscillator circuits is to add a third variable capacitance in parallel with the fixed series capacitances, as shown in Fig. 2-7. This technique, by definition, keeps the $C_1{:}C_2$ ratio constant, because both of these capacitors have fixed values, but the overall resonant frequency is variable. The total effective capacitance in the tank is dependent on all three capacitors—C1, C2, and C3. Remember, capacitances in parallel add together. Therefore, the formula for the total effective tank capacitance is:

$$C_t = C_3 + \left(\frac{(C_1 \times C_2)}{(C_1 + C_2)} \right)$$

The Colpitts oscillator is a very popular circuit, because it offers very good frequency stability at a reasonable cost.

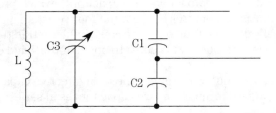

2-7 A simple way to make a Colpitts oscillator circuit's frequency variable over a small range is to add a third variable capacitance in parallel with the fixed series capacitances.

The ultra-audion oscillator

As might be expected with any popular circuit type, a number of variations on the basic Colpitts oscillator circuit have been devised over the years. One of the most frequently encountered variants is the *ultra-audion oscillator*. A typical circuit of this type is shown in Fig. 2-8. While our main interest in this chapter is on audio-frequency oscillators, the ultra-audion oscillator circuit is very well-suited for use in the VHF (very high frequency) range. This type of circuit is often used as the local oscillator in television receivers.

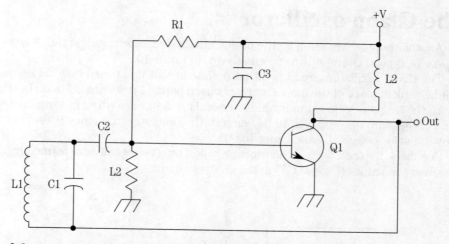

2-8 A popular variation on the basic Colpitts oscillator circuit is the ultra-audion oscillator circuit.

Internal capacitances within the transistor itself provide feedback paths at VHF frequencies. These internal capacitances function just as if they were small external capacitors, connected from the base to the emitter, and from the emitter to the collector, as illustrated in Fig. 2-9. Because of their extremely small size, these internal capacitances have very low reactances in the VHF range.

The ultra-audion oscillator circuit can also be used at lower frequencies, but larger external capacitors will need to be added to take the place of the too-small internal capacitances within the transistor.

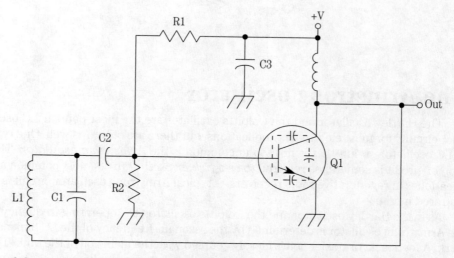

2-9 Internal capacitances within the transistor itself provide feedback paths at VHF frequencies.

The Clapp oscillator

Another popular variation on the basic Colpitts oscillator circuit is the *Clapp oscillator*. A typical circuit of this type is shown in Fig. 2-10.

The Clapp oscillator circuit is rather unique in that it is tuned by a series-resonant LC tank in place of the more commonly used parallel-resonant LC tank. In other words, the frequency-determining inductor and capacitor are wired in series instead of in parallel with one another. In this circuit, the oscillator frequency is set primarily by the values of coil L1 and capacitor C1.

Feedback for the Clapp oscillator is provided by a voltage divider formed by the two small capacitors (C2 and C3) in the emitter circuit.

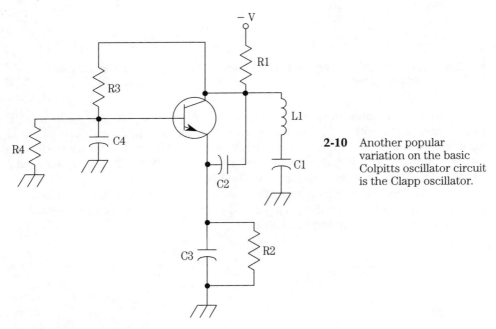

2-10 Another popular variation on the basic Colpitts oscillator circuit is the Clapp oscillator.

The Armstrong oscillator

The Hartley oscillator and the Colpitts oscillator are the most popular LC oscillator circuits for most electronics applications, but there are others as well. One type of LC oscillator circuit that is not too uncommon is the *Armstrong oscillator*. This type of circuit is sometimes referred to as a *tickler oscillator*. Its most common application is in regenerative radio receivers. A typical Armstrong oscillator circuit is illustrated in Fig. 2-11.

As in the Hartley oscillator and the Colpitts oscillator, the operating frequency of the Armstrong oscillator is determined by the resonant frequency of the LC tank network. A feedback, or tickler coil is closely coupled with the main coil of the tank. The tickler coil feeds a small part of the output signal back to the amplifier's (Q1's) input.

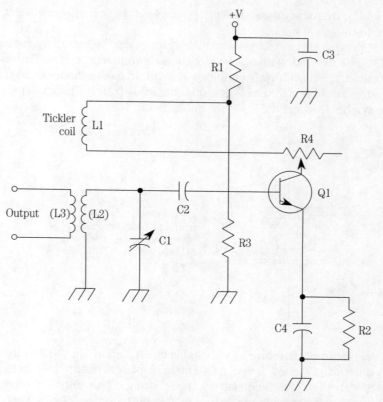

2-11 The Armstrong oscillator circuit is often used in regenerative radio receivers.

These coils must be very carefully positioned so that their mutual inductance is of the proper polarity, or the circuit might not oscillate reliably, if at all.

Potentiometer R4 controls the level of the current flowing in the tickler coil, and thus, the amount of feedback or regeneration. This potentiometer is omitted in some practical Armstrong oscillator circuits.

The crystal oscillator

A discrete coil and capacitor tank is not the only possible way to set up a resonant frequency within an electronic circuit. It is generally the least expensive, and most convenient way—but it isn't necessarily the best way. For one thing, practical coil and capacitor values are often a little less than precise. These components commonly have fairly wide tolerances. Moreover, their actual values can be temperature-sensitive, or could vary in response to other environmental or circuit conditions. The resonant frequency also tends to drift under some circumstances. Often this type of error is insignificant, which is why LC-based oscillator circuits are so popularly employed throughout the field of electronics. But in some applications demanding high precision and/or frequency consistency, the simple LC tank simply might not be good enough to do the job at hand.

A more precise and reliable resonant frequency can be set up by replacing the discrete LC tank network with a specially cut slab of quartz crystal. This new component is called, unimaginatively enough, a *crystal*. In most schematic diagrams and parts lists, this is abbreviated as XTAL, or sometimes (rarely) just X. The standard schematic symbol for a crystal is shown in Fig. 2-12.

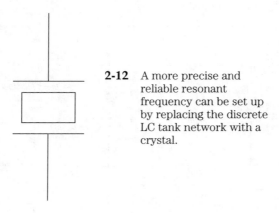

2-12 A more precise and reliable resonant frequency can be set up by replacing the discrete LC tank network with a crystal.

The basic internal structure of a crystal is illustrated in Fig. 2-13. A thin slice of quartz crystal is sandwiched between two metallic plates that are held in tight physical contact with the crystalline slab by small springs. This entire assembly is enclosed in a hermetically-sealed metal case. The hermetic seal helps keep out any potentially harmful contamination, such as moisture or dust, which could eat away at the delicate crystal slab and/or alter its electrical value. Leads connected to the individual metal endplates are brought out from the bottom of the case to permit connection to external circuitry.

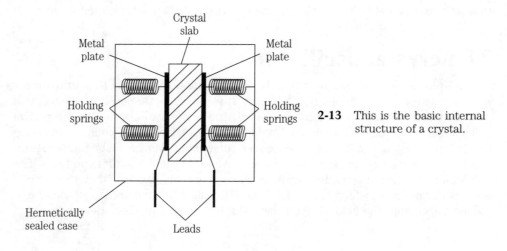

2-13 This is the basic internal structure of a crystal.

A crystal works because of a special phenomenon known as the piezoelectric effect. Two sets of axes pass through the body of any crystal. One set, called the *X axis*, passes through the corners of the crystal. The other set, called the *Y axis*, lies perpendicular (at a 90-degree angle) to the X axis, but in the same plane, as illustrated in Fig. 2-14. Practical crystals for electronic use are made of a very thin slice of the crystalline material (almost always quartz). This crystal slice can be cut along either an X axis or a Y axis.

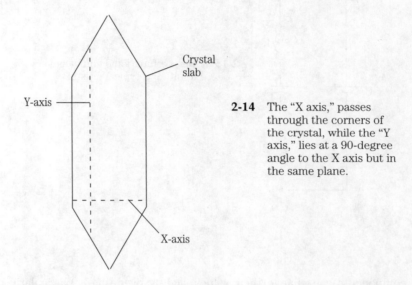

Crystal
slab

Y-axis

X-axis

2-14 The "X axis," passes through the corners of the crystal, while the "Y axis," lies at a 90-degree angle to the X axis but in the same plane.

If a mechanical stress is placed across the Y axis of a crystal, an electrical voltage will be produced across the X axis. Similarly, if an external electrical voltage is applied across the X axis, an internal mechanical stress will appear along the corresponding Y axis. This is the *piezoelectric effect*. In practice, the piezoelectric effect can cause the crystal to ring (or vibrate) at a specific rate—that is, it resonates at a specific frequency under certain controlled conditions.

Electrically, a crystal is a somewhat complex device. Basically, it functions as if it were made up of the circuit shown in Fig. 2-15. Depending on how the crystal is manufactured, it can be used as either a series-resonant, or a parallel-resonant LC tank network. Generally speaking, a crystal designed for series-resonant use cannot be used in a parallel-resonant circuit, and vice versa.

The resonant frequency of a specific crystal is determined primarily by the thickness and size of the crystal slice. Under some circumstances, crystals can also be made to resonate at integer multiples (harmonics of its main resonant frequency. For example, a crystal cut for 1.5 MHz, can be forced to resonate at 3.0 MHz (2×1.5 MHz), or 4.5 MHz (3×4.5 MHz). However, the resonance effect becomes increasingly less pronounced at higher harmonics.

In most cases, crystal is more expensive than separate capacitors and coils. Also, it is not easy to alter the resonant frequency in a crystal-based oscillator circuit. On the other hand, LC-resonant circuits can drift off frequency (that is, the frequency deter-

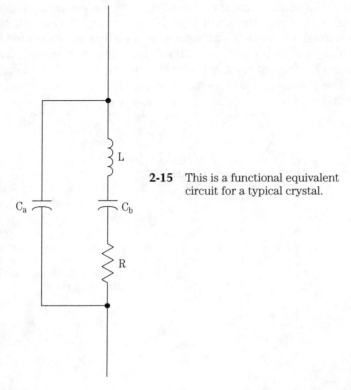

2-15 This is a functional equivalent circuit for a typical crystal.

mining components can change their values slightly over time), particularly under changing temperature conditions, like when a circuit heats up under continuous usage.

Crystals are also somewhat temperature-sensitive, although not as much so as most practical capacitors and coils. When extremely high accuracy is required (as in broadcasting applications, for example), a crystal oscillator circuit is usually enclosed in a special crystal oven, which maintains a constant temperature environment for the crystal.

Reliability is another major advantage of crystals. The failure rate for crystals is significantly lower than for discrete capacitors and coils. That is, they need to be replaced for repair less often. However, a crystal can be damaged by high over-voltages, or by extremely high temperatures. A severe mechanical shock (such as being dropped some distance onto a hard surface) can crack the delicate crystal slice, causing it to operate erratically, or not operate at all.

The most important disadvantage of crystals is the same as their most important advantage. The resonant frequency of the crystal is inherent in its manufacture, and is a constant. The resonant frequency can not be easily changed while the oscillator circuit is in operation. In a LC circuit, one or both of the frequency-determining components can be made variable. But there is no such thing as a variable crystal.

Occasionally, special external circuitry can be added to a crystal oscillator to permit a limited degree of fine tuning, but the range will be very small, and such techniques are the exception, rather than the rule.

The only way the frequency of a crystal-based circuit can be changed is to physically replace the crystal itself. This is why crystals are usually inserted into sockets, rather than permanently soldered into the circuit. Crystals can easily and quickly be removed and replaced (with the circuit's power disconnected, of course). Using sockets also sidesteps the potential problem of thermal damage that can result from repeatedly soldering and desoldering the crystal's leads. In some circuits, multiple crystals are wired in parallel, and can be selected by a multi-position switch.

There are many different types of crystal oscillator circuits. Basically, a crystal can be used in place of almost any standard LC-resonant tank network. A fairly typical parallel-resonant crystal oscillator circuit is shown in Fig. 2-16. Figure 2-17 shows a typical series-resonant crystal oscillator circuit.

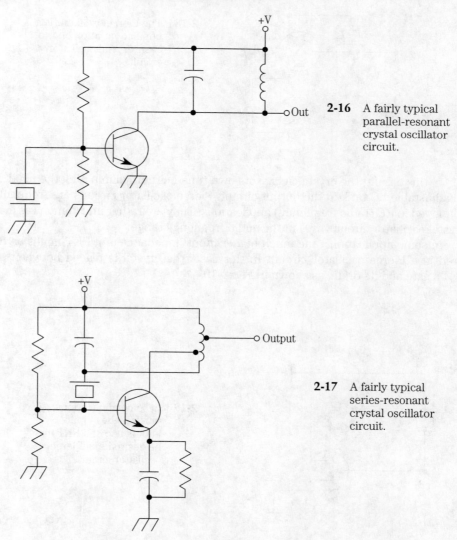

2-16 A fairly typical parallel-resonant crystal oscillator circuit.

2-17 A fairly typical series-resonant crystal oscillator circuit.

The Pierce oscillator

The *Pierce oscillator* is a popular variation on the basic crystal oscillator circuit. In this type of circuit, the crystal is placed between the base and the collector of the amplifier transistor, as illustrated in Fig. 2-18.

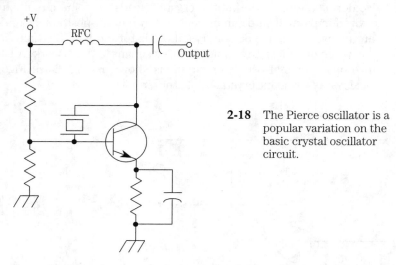

2-18 The Pierce oscillator is a popular variation on the basic crystal oscillator circuit.

In this circuit, the crystal acts as its own tuned circuit, eliminating the need for an adjustable LC tank in the output circuit. Pierce oscillator circuits are frequently employed in RF (radio frequency) applications. They are rather uncommon (as most crystal oscillator circuits are) in the audio frequency range.

In some applications, the much higher input impedance of a FET might be desired in a Pierce oscillator circuit. In this case, the crystal is placed between the FET's gate and its drain, as shown in Fig. 2-19.

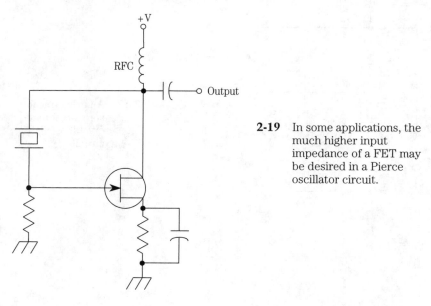

2-19 In some applications, the much higher input impedance of a FET may be desired in a Pierce oscillator circuit.

Op amp oscillators

In the last few decades, more and more electronic circuits have appeared in integrated circuit form. So far, we have only looked at basic sine-wave oscillator circuits using discrete components.

Probably the most popular single type of IC is the *operational amplifier*, or *op amp*. We will be using this device in many of the circuits throughout the book. Sometimes, it seems like the op amp can be used in almost any electronics application (although that isn't quite the case). Op amps can be used to generate very pure, essentially distortion-free sine waves at reasonably low expense.

One of the most commonly used op amp sine-wave oscillator circuits is illustrated in Fig. 2-20. This circuit is often called a *twin-T oscillator*.

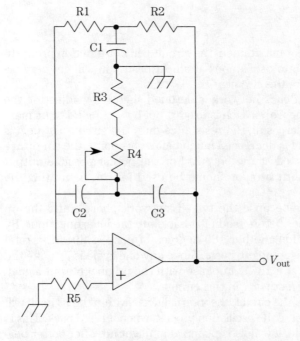

2-20 The "twin-T" oscillator circuit is one of the most commonly used op amp sine-wave oscillator circuits.

The secret to the operation of this circuit lies in its special feedback network, the twin-T network. Similar networks are often used in filter circuits, and several other common electronics applications. A *twin-T* is obviously made up of two Ts. This is not an esoteric technical term, but just a description of the way the circuit looks in most schematic diagrams. Resistors R1 and R2, along with capacitor C1, form one T. The other T is made up of resistors R3 and R4, and capacitors C2 and C3. In this particular schematic, the second T is drawn upside down, for convenience.

The formula for the oscillation frequency in this circuit is:

$$F = \frac{1}{(2 \pi R_1 C_2)}$$

Since pi (π) is a mathematical constant that always has a fixed value of about 3.14, we can rewrite this formula as:

$$F = \frac{1}{(6.28\,R_1 C_2)}$$

This formula assumes that the component values throughout the circuit have the following important relationships:

$$R_2 = R_1$$

$$R_3 = \frac{R_1}{4}$$

$$R_4 = \frac{R_1}{2} \text{ (approximate)}$$

$$C_1 = 2C_2$$

$$C_3 = C_2$$

If these value relationships are not maintained, the circuit will not function properly, and might not be able to break into oscillations at all. At best, you will get very erratic, hard-to-predict results from the circuit.

In operation, the twin-T feedback network is detuned slightly by adjusting the value of potentiometer R4. The best approach is to start out by setting R4 to its maximum resistance, then slowly decreasing the resistance until the circuit just begins to oscillate. If the resistance of R4 is decreased much below this point, the sine-wave signal will be increasingly distorted at the output. For the purest possible output waveform, a high-precision 10-turn trimpot should be used for R4, permitting the most accurate setting possible.

This oscillator circuit works because the twin-T network phase-shifts the op amp's output signal by 180 degrees before feeding it back into the inverting input. By definition, the inverting input adds another 180 degrees of phase shift, for a total phase shift of 360 degrees, or one complete cycle. Since any one cycle is just like the next one, the feedback signal is effectively in phase with the original output signal. In other words, there is *positive feedback* in this circuit.

When power is first applied to the circuit, the internal noise within the IC itself will generate enough of a signal to initiate the oscillation process when it is fed back. This is one application where a high-grade, low-noise op amp chip might not offer better performance. In fact, it might not work at all. Unless you are dealing with very high frequencies, well above the limits of the audio range, a standard, inexpensive, 741—or one of its close relatives—is usually the most appropriate choice for this application.

In practical circuit design work, you will presumedly know the desired output signal frequency, and will need to find the necessary component values. This is not hard to do. As an example, we will quickly run through the design of a twin-T oscillator circuit with an output signal frequency of 1,700 Hz (1.7 kHz). The first step is to algebraically rearrange the basic frequency formula:

$$R_1 = \frac{1}{(6.28 F C_2)}$$

Now, select a likely value for capacitor C2. This is somewhat of an arbitrary choice. If you get awkward resistor values from the following calculations, just pick a new value for capacitor C2, and try again. As you get used to working with such formulae, you will get a good "feel" for likely capacitor values for almost any desired frequency range.

In this case, let's try using a 0.022-μF capacitor for C2. This means the required value for resistor R1 works out to about:

$$R_1 = \frac{1}{(6.28 \times 1200 \times 0.000000022)}$$

$$= \frac{1}{0.0002348}$$

$$= 4{,}259 \text{ ohms}$$

Unless our application is very critical about the exact frequency, we can use a standard 5% tolerance, 4.2-kΩ (4,200 ohms) resistor for R1.

The hard part of the design is done already. Now, we just have to set up the other component value relationships throughout the circuit. First, the resistors:

$$R_2 = R_1$$

$$= 4.2 \text{ k}\Omega$$

$$R_3 = \frac{R_1}{4}$$

$$= \frac{4200}{4}$$

$$= 1050 \text{ ohms}$$

A standard 1-kΩ resistor can be used for R3. The value of this resistor will always be less critical, and a precision resistor will rarely be needed, because it is in series with potentiometer R4, which can be adjusted to make up for any inaccuracy in R3's value:

$$R_4 = \frac{R_1}{2}$$

$$= \frac{4200}{2}$$

$$= 2100 \text{ ohms}$$

A 2.5-kΩ potentiometer would be a good choice for R4 in our example circuit. Again, because the potentiometer's resistance is adjustable, you don't have to worry about finding a component that has a rating exactly equal to the calculated value. In fact, it is a very good idea to use a potentiometer that can cover resistances a little higher than the calculated value, to leave some headroom for compensating for other inaccuracies throughout the circuit.

Finding the rest of the capacitor values is even easier:

$$C_3 = C_2$$

$$= 0.022 \text{ μF}$$

$$C_1 = 2C_2$$
$$= 2 \times 0.022$$
$$= 0.044 \ \mu F$$

A standard 0.047-μF capacitor should be close enough for C1. Remember, most practical capacitors have relatively wide tolerances.

In some cases, you might find equations that call for a really odd-ball value for C1. In such cases, it is easy enough to use two paralleled capacitors with the same value as C2 and C1. Remember, capacitances in parallel add, so this will always give us the correct value (within the limits of the component tolerances, of course). And that is all there is to designing this circuit.

Nothing is perfect, and this twin-T oscillator circuit is no exception to that rule. The major disadvantage of this circuit is that the frequency and the purity of the output waveform is dependent on the interrelated values of all of the passive components. There is no easy way to make the output frequency-variable.

If your intended application demands a variable-output frequency, a *quadrature oscillator* circuit might be a better choice. A typical circuit of this type is illustrated in Fig. 2-21. The chief disadvantage of this circuit is that it requires two op amps, so it will be more expensive than the twin-T circuit.

A quadrature oscillator has two outputs, which are labelled *sine* and *cosine*. A cosine wave is simply a sine wave that has been phase-shifted 90 degrees, as illustrated in Fig. 2-22. In many practical applications, the cosine output will not be needed. That's OK. Just ignore it and don't connect it to anything. No law says that you have to use every available output from a given circuit.

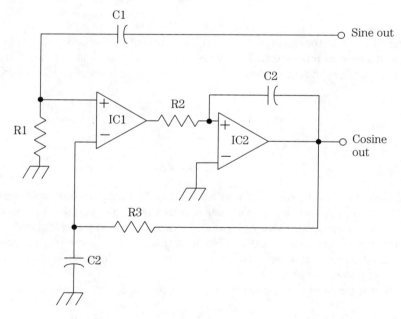

2-21 A quadrature oscillator is a better choice when a variable-output frequency is required from an op amp sine-wave oscillator circuit.

Sine

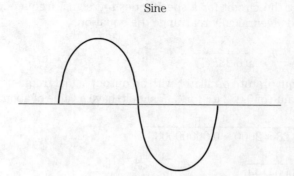

2-22 A cosine wave is simply a sine wave that has been phase-shifted 90 degrees.

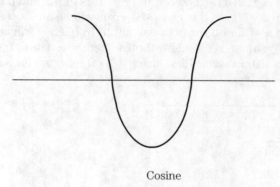

Cosine

This basic quadrature oscillator circuit uses only six passive components in addition to the two op amps, so there won't be many design equations. The design process is even simpler than it appears at first glance, because all three capacitors (C1, C2, and C3) have identical values. Similarly, resistors R2 and R3 are also equal in value. Resistor R1 is given a value slightly less than that of R2 to make sure the oscillations will begin reliably as soon as power is applied to the circuit. The exact value is not particularly critical here. Generally, the most convenient approach is to use the next lower standard resistor value for R1, after calculating the value of R2. There is no need to perform any additional calculations for R1.

All of this means that we really need to determine just two component values when designing this circuit for a specific output frequency—C (C_1, C_2, and C_3) and R (R_2 and R_3).

The frequency equation of this quadrature oscillator circuit is the same as for the twin-T oscillator circuit discussed above:

$$F = \frac{1}{(6.28RC)}$$

Once again, when designing the circuit for a specific desired signal frequency, just pick a likely value for C, and algebraically rearrange the equation:

$$R = \frac{1}{(6.28FC)}$$

For example, let's design a quadrature oscillator with an output signal frequency of 2000 Hz (2 kHz). If we use a value of 0.033 µF for C, R should have a value of about:

$$R = \frac{1}{(6.28 \times 2000 \times 0.000000033)}$$

$$= \frac{1}{0.0004144}$$

$$= 2{,}413 \text{ ohms}$$

We could use 2.7-kΩ (2,700 ohms) resistors for R2 and R3, and a 2.2-kΩ (2,200 ohms) resistor for R1—or we could use 2.2-kΩ (2,200 ohms) resistors for R2 and R3, and a 1.8-kΩ (1,800 ohms) resistor for R1. In this case, the error will be about the same either way. If you want to see how far off the nominal output frequency will be (ignoring component tolerances), you can recalculate the frequency equation for each of the two possible standard resistor values. First, using 2.7-kΩ resistors for R2 and R3, the nominal output frequency works out to:

$$R = \frac{1}{(6.28 \times 2700 \times 0.000000033)}$$

$$= \frac{1}{0.0005595}$$

$$= 1{,}787 \text{ Hz}$$

A little low (from the desired 2,000 Hz). Now, we can try using 2.2 kΩ resistors for R2 and R3. This will give us a nominal output frequency of about:

$$R = \frac{1}{(6.28 \times 2200 \times 0.000000033)}$$

$$= \frac{1}{0.0004559}$$

$$= 2{,}193 \text{ Hz}$$

This time, the nominal frequency comes out a bit higher than we originally wanted. Remember, component tolerances are likely to account for at least as much error in either direction. In most applications, such variations shouldn't matter too much. In critical applications, requiring more precise output frequencies, it will probably be necessary to use high-grade, low-tolerance components throughout the circuit.

Besides the basic simplicity and elegance of its design, the quadrature oscillator also offers the advantage of a potentially variable-output frequency—at least over a limited range. This can easily be accomplished by replacing R2 and R3 with a dual potentiometer. The shafts are ganged together, so these two resistances will always be equal, as required by the circuit. Fixed resistors should probably be used in series with the potentiometers, so that the value of R never drops below that of resistor R1.

Also, the circuit might not function properly if the value of R is too much greater than that of R1. This is why the signal frequency can usually be adjusted only over a relatively narrow range in this circuit.

The sine-wave and cosine-wave outputs from this quadrature oscillator circuit will always have the exact same frequency. The only difference between the two output signals is their relative phase. The two outputs can be used independently or together, depending on the application. In many practical applications, the cosine output is simply left unused.

Dedicated ICs

Dedicated ICs have been developed for most common (and many rather uncommon) electronics functions. A number of oscillator chips exist, but most can be use for other applications and are not strictly sine-wave oscillators. For example, there are function-generator ICs like the 8038 and the XR2206. They can be used as sine-wave oscillators, but they can also generate other waveforms. These devices will be discussed in more detail in Chapter 4.

We will use the XR2206 function-generator IC in the audio oscillator project for this chapter. The pin-out diagram for this device is shown in Fig. 2-23. This chip is manufactured by Exar. I have seen it second-sourced from other manufacturers, with different prefix letters. (XR is used exclusively by Exar). At the time of this writing, the XR2206 is reasonably well-available to the electronics hobbyist, but always remember that electronics is a rapidly changing field, and there is no way for

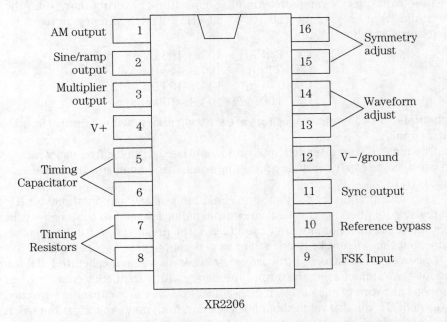

XR2206

2-23 The XR2206 function generator IC can generate the popular electronic waveforms, including the sine wave.

the author or the publisher to guarantee any particular device won't suddenly be discontinued and become obsolete and impossible to find. Always make sure you can find a source for any specialized ICs (and any other unusual components) before spending any money on parts for any electronics project. You can minimize your disappointment and frustration if you find out a critical component is no longer available before you spend a lot of time and money gathering the rest of the parts for a project you can't complete because of an unanticipated change in the electronics marketplace. I expect the XR2206 to be around for years to come, but there is no way for me to guarantee this.

Project #3—Sine-wave audio oscillator

A practical circuit for generating very pure sine waves over a wide frequency range is shown in Fig. 2-24. A suitable parts list for this project is given in Table 2-1.

This sine-wave audio oscillator project is built around the XR2206 function-generator IC. This IC can be operated over a 2000:1 frequency range. When properly calibrated, the sine-wave output signal can have less than 2.5% distortion—which is pretty good, considering the low total cost of this project.

The XR2206's internal design permits adjustment over its full frequency range with a single external potentiometer. But unfortunately, such a wide range control can be difficult to adjust precisely. For many testing procedures, a technician needs a very specific signal frequency. To make fine tuning the circuit easier, this project features a range selector switch (S1), that permits you to choose between four frequency-range-determining capacitors, covering more than the entire audio frequency range (20 Hz to 20 kHz). The frequency range for each capacitor is as follows:

C4	1 μF	10 Hz–100 Hz
C5	0.1 μF	100 Hz–1 kHz
C6	0.01 μF	1 kHz–10 kHz
C7	0.001 μF	10 kHz–100 kHz

High-grade, low-tolerance capacitors are strongly recommended for use in this part of the circuit.

Of course, if your intended application only needs a single frequency range, you can use a single capacitor of an appropriate value, and eliminate the range selector switch (S1).

The actual frequency within the selected range is set via potentiometer R1. Series resistor R2 prevents this resistance from being adjusted to too small a value. A precision multi-turn potentiometer would give the greatest fine-tuning accuracy, of course, but this will not be needed in most general-purpose applications.

Once you've constructed the project, you can make up a calibrated dial for potentiometer R1 monitoring the output frequency with a frequency counter or an oscilloscope at a variety of settings. If high-grade, low-tolerance capacitors are used for C4 through C7, the dial value should be constant from range to range. You will need just a single set of calibration marks, and will only need to mentally move the decimal point according to the currently selected frequency range.

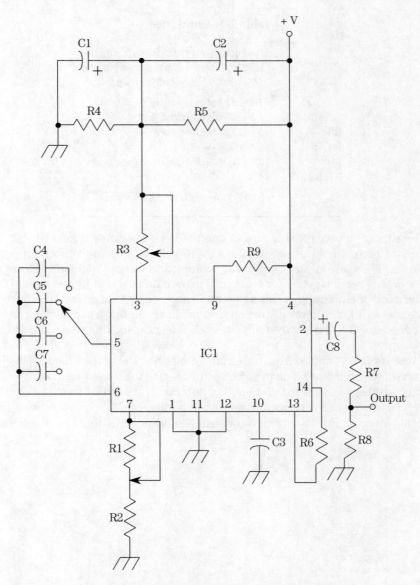

2-24   Project #3—Sine-wave audio oscillator.

**Table 2-1. Suggested parts list for Project #3—
Sine-wave audio oscillator of Fig. 2-24**

IC1	XR-2206 function-generator IC
C1, C2, C3	10-µF, 35-V electrolytic capacitor
C4	1-µF, 35-V tantalum capacitor
C5	0.1-µF mylar capacitor, or similar

Table 2-1. Continued

C6	0.01-μF mylar capacitor, or similar
C7	0.001-μF mylar capacitor, or similar
C8	100-μF, 35-V electrolytic capacitor
R1	100-kΩ potentiometer (see text)
R2, R9	10-k, ¼-W, 5% resistor
R3	50-kΩ potentiometer
R4, R5	5.6-kΩ, ¼-W, 5% resistor
R6	220-Ω, ¼-W, 5% resistor (see text)
R7, R8	470-Ω, ¼-W, 5% resistor
S1	SP4T four-position rotary switch

The supply voltage for this project can theoretically range from +10 volts up to +26 volts. I recommend using a +12-volt, +15-volt, or +18-volt power supply with this circuit. These are all standard voltage values, and voltage-regulator ICs are available for each. (See chapter 1 for more information on power-supply circuits.)

The gain, or amplitude of this audio oscillator's output signal can be adjusted via potentiometer R3. For lower distortion, you might try adding a calibration trimpot in place of resistor R6, and slowly adjust it for the cleanest-looking sine wave on a good oscilloscope.

If resistor R6 is omitted from the circuit altogether, the output signal will be a triangle wave, which might be an even better test signal for some purposes. You might want to consider adding a switch to add or subtract this resistor from the circuit. Of course, this modification turns our sine-wave audio oscillator project into a sort of function generator. Function generators will be discussed in detail in chapter 4.

3
CHAPTER

Rectangle-wave, square-wave, and pulse-wave generators

In the early days of electronics, the most commonly-used ac waveform was the sine wave—and why not? In its simplicity, it was the most obvious and natural choice in analog circuits. In recent years, however, the rectangle wave (and its variants) has more than surpassed the old reliable sine wave in terms of popularity in usage. One reason for the widespread use of rectangle waves in modern electronics is that it is the one ac waveform that is suitable for use in both analog and digital circuits.

In the course of its cycle, a sine wave slides smoothly from value to value. (A sine wave is illustrated again in Fig. 3-1, for your convenience.) It starts out at zero, builds smoothly up to a positive-peak value, then immediately changes direction (in a smooth curve, rather than an abrupt switch), and starts to smoothly decrease in value, through zero, and on down to a negative-peak value (usually, though not always equal to the positive-peak value, except for the reversed polarity). Again, the signal goes through a smooth, even curve to change direction, and starts to increase in value, up to the zero line, where a new cycle begins, and the entire process repeats.

Notice that no single value is ever held more for the tiniest fraction of an instant. The voltage (or current) is constantly varying, and never stays still. The signal constantly passes through all of the intermediate values between the positive and negative peak values. There are an infinite number of instantaneous values in this (and almost any other) ac waveform.

A rectangle wave, on the other hand, is made up of just two discrete values. The signal switches back and forth between these two values, theoretically with no transition between them. The signal's instantaneous value is either HIGH (positive peak), or LOW (negative peak), and never anything in between. A typical rectangle wave is shown in Fig. 3-2.

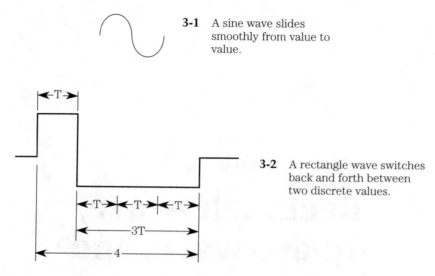

3-1 A sine wave slides smoothly from value to value.

3-2 A rectangle wave switches back and forth between two discrete values.

Of course, no practical circuitry can switch between widely-spaced values truly instantaneously. There will be some finite transition time, causing a slant in the sides of the waveform. This is illustrated in very exaggerated fashion in Fig. 3-3. This effect is known as *slew*. The slew rate is an important specification for rectangle-wave signal generators and switching circuits. The smaller the slew rate value, the more truly square the waveform is. Relatively simple and inexpensive signal generator circuits can produce rectangle waves with very good slew rates, so this is rarely a significant problem—except in the most critical high-precision applications.

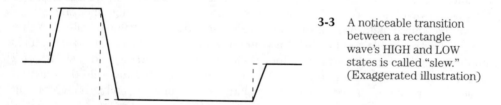

3-3 A noticeable transition between a rectangle wave's HIGH and LOW states is called "slew." (Exaggerated illustration)

In most cases, the transition between the LOW and HIGH states comprises considerably less than 1% of the total cycle time, and it can reasonably be ignored in most practical work.

Duty cycle

A sine wave is pretty much a sine wave. Except for distortion effects, which would make the signal less of a sine wave, not much can change except the signal frequency and/or the overall amplitude. But there are many different types of rectangle waves. Two specialized forms of rectangle waves have their own names—*square waves* and *pulse waves*. Often, these terms are used interchangeably. Some writers might refer to any rectangle wave as a square wave, or might call a square wave a type of pulse wave. Such usage isn't quite correct. The three terms exist for a reason. They are not merely redundant. However, the use of the term pulse wave to refer to any rectangle wave is fairly widespread today, especially when talking about signals in digital circuitry. We will get to the true definitions of square wave, and pulse wave shortly.

The difference between various types of rectangle waves is defined by a specification known as the *duty cycle*. This is a ratio of the HIGH time per cycle to the total cycle time, in this form:

$$T_h{:}T_t$$

where T_h is how long per cycle the signal is held at the HIGH state, and T_t is the length of the entire cycle (both HIGH and LOW times). The first number in the duty cycle is almost always given as a 1, with the second number (the total cycle time) adjusted accordingly. Remember, it is the ratio that is of importance here, not the absolute time values. The duty cycle is not normally dependent on the signal frequency (except in certain circuits where absolute times are manipulated for various reasons of circuit design).

The rectangle wave shown back in Fig. 3-2 has a duty cycle of 1:4. The signal is HIGH for ¼ of each complete cycle. Figure 3-4 shows a couple of additional rectangle waves, with different duty cycles.

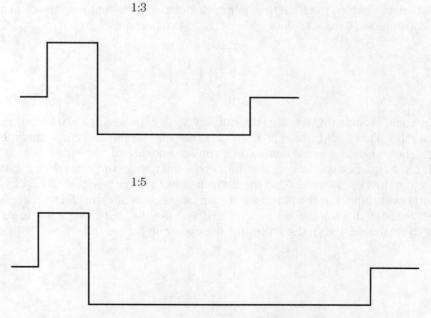

3-4   A rectangle wave can have many different duty cycles.

Let's imagine a 100-Hz rectangle wave. The total cycle time of any ac signal is equal to the reciprocal of the frequency, so:

$$T_t = \frac{1}{F}$$

$$= \frac{1}{100}$$

$$= 0.01 \text{ second}$$

$$= 10 \text{ mS}$$

If the signal is HIGH for 2.5 mS per cycle, the duty cycle of the signal is:

$$dc = T_h : T_t$$

$$= 2.5 : 10$$

$$= \left(\frac{2.5}{2.5}\right) : \left(\frac{10}{2.5}\right)$$

$$= 1 : 4$$

On the other hand, if the signal is held HIGH for 3.33 mS per cycle, the duty cycle works out to:

$$dc = 3.33 : 10$$

$$= \left(\frac{3.33}{3.33}\right) : \left(\frac{10}{3.33}\right)$$

$$= 1 : 3$$

When we shorten the HIGH time per cycle down to 2 mS, but hold the total cycle time constant at 10 mS, the duty cycle of the signal becomes:

$$dc = 2 : 10$$

$$= \left(\frac{2}{2}\right) : \left(\frac{10}{2}\right)$$

$$= 1 : 5$$

In some technical literature, the duty cycle is expressed as a fraction, rather than a ratio. That is, ⅕ instead of 1:5. This is not inaccurate, and can sometimes be helpful, but it can also cause confusion if you are not careful.

If the frequency is changed, the duty cycle will remain the same. In our last example, the duty cycle was 1:5, so the HIGH time will always be equal to exactly ⅕ of the total cycle time. For the purposes of calculations, it is often useful to express the duty cycle as such a fraction. For example, if we raise the signal frequency to 200 Hz (total cycle time = 5 mS), the HIGH time per cycle will be:

$$T_h = T_t \times dc$$

$$= 5 \times \frac{1}{5}$$

$$= \frac{5}{5}$$

$$= 1 \text{ mS}$$

Similarly, if we decrease the signal frequency in this example down to 50 Hz (T_t = 20 mS), the HIGH time per cycle works out to:

$$T_h = 20 \times \frac{1}{5}$$

$$= \frac{20}{5}$$

$$= 4 \text{ mS}$$

A very important special case of the rectangle wave is when the HIGH time per cycle is exactly half of the total cycle time. That is, the duty cycle is 1:2. Such a waveform is illustrated in Fig. 3-5. This is a true square wave. To be completely accurate, a square wave always has a duty cycle of 1:2. If it has any other duty cycle value, it is not really a square wave. All square waves are rectangle waves, but not all rectangle waves are square waves.

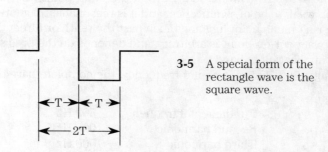

3-5 A special form of the rectangle wave is the square wave.

A true pulse wave goes to the opposite extreme. The signal is HIGH for only a very small fraction of the total cycle time. For example, typical pulse waves might have duty cycles of 1:10 or 1:24, or even higher ratios.

There is no hard and fast rule of just when a rectangle wave becomes a pulse wave. Some technical writers call all rectangle waves pulse waves, especially when speaking of such signals in digital systems. Personally, I think it makes sense to consider a rectangle wave a pulse wave only if the HIGH time per cycle is ⅒ or less of the total cycle time.

Harmonics

In the last chapter, I mentioned that the simple sine wave consists of just a single frequency component, the fundamental, but all other ac waveforms are more complex, with multiple frequency components. The fundamental is usually the lowest frequency component, and the one with the highest amplitude. It is the base frequency of the waveform as a whole. When we say "such-and-such a waveform has a frequency of 500 Hz," we mean its fundamental frequency is 500 Hz. Additional frequency components (almost always with progressively weaker amplitudes as they get further from the fundamental frequency) are known as *overtones*. For a repeating waveform, like a rectangle wave, the overtones are always exact whole-number multiples of the fundamental frequency. In this case, they are known as *harmonics*. That is, a harmonic overtone can be 2 times the fundamental frequency or 3 times the fundamental frequency, but not 2.5 times the fundamental frequency.

The multiple value determines the harmonic number. The harmonic that is 2 times the fundamental frequency is the second harmonic, the one at 3 times the fundamental frequency is the third harmonic, and so forth. There is no harmonic below the second. If you want, you can consider the fundamental frequency itself as the first harmonic, because its frequency is, by definition, equal to 1 times the fundamental frequency.

The fundamental frequency, as well as each harmonic frequency, can be considered to be individual sine waves at the appropriate frequencies and amplitudes, that are blended together to make up the complex waveform. Building up a complex waveform from individual sine waves is called *additive synthesis*.

In most ac waveforms (including all of the commonly used ones), the higher the harmonic, the weaker its amplitude. Beyond a certain point, the harmonic amplitudes are too weak to be of significance, and it is reasonable to ignore them. In most cases, there isn't much point in going much past the tenth or fifteenth harmonic, although the exact cut-off point is arbitrary, and depends on the precision required in the specific application at hand.

Assuming a 500-Hz fundamental frequency, the complete harmonic series is as follows:

Fundamental frequency	500 Hz
Second harmonic	1000 Hz
Third harmonic	1500 Hz
Fourth harmonic	2000 Hz
Fifth harmonic	2500 Hz
Sixth harmonic	3000 Hz
Seventh harmonic	3500 Hz
Eighth harmonic	4000 Hz
Ninth harmonic	4500 Hz
Tenth harmonic	5000 Hz
Eleventh harmonic	5500 Hz
Twelfth harmonic	6000 Hz

and so forth.

Of course, increasing the fundamental frequency will cause all of the harmonics to increase in frequency by a like amount. The same thing happens if the fundamental frequency is decreased.

Not every complex ac waveform includes every possible harmonic. In fact, the pattern of harmonics and their relative amplitudes are the primary factors in determining the waveshape.

In a rectangle wave, the second number of the duty cycle (representing the proportion of the total cycle time) determines which harmonics are omitted from the complete pattern. For example, a rectangle wave with a duty cycle of 1:3 would include all harmonics, except for those that are exact multiples of 3. Assuming the fundamental frequency is still 500 Hz, the make-up of this signal will look like this:

Fundamental frequency	500 Hz
Second harmonic	1000 Hz
Fourth harmonic	2000 Hz
Fifth harmonic	2500 Hz
Seventh harmonic	3500 Hz
Eighth harmonic	4000 Hz
Tenth harmonic	5000 Hz
Eleventh harmonic	5500 Hz

and so forth.

If the rectangle wave has a duty cycle of 1:4, every fourth harmonic is omitted from the signal's harmonic make-up:

Fundamental frequency	500 Hz
Second harmonic	1000 Hz
Third harmonic	1500 Hz
Fifth harmonic	2500 Hz
Sixth harmonic	3000 Hz
Seventh harmonic	3500 Hz
Ninth harmonic	4500 Hz
Tenth harmonic	5000 Hz
Eleventh harmonic	5500 Hz

and so forth.

A square wave always has a duty cycle of 1:2. It includes the fundamental frequency and all of the odd harmonics, but none of the even harmonics (multiples of 2):

Fundamental frequency	500 Hz
Third harmonic	1500 Hz
Fifth harmonic	2500 Hz
Seventh harmonic	3500 Hz
Ninth harmonic	4500 Hz
Eleventh harmonic	5500 Hz

and so forth.

As you can imagine, true pulse waves are very rich in harmonic content, because the first harmonic omitted is so far up the series.

In a rectangle wave, the relative amplitude of each harmonic frequency component is equal to the fundamental frequency divided by the harmonic number. That is:

Fundamental frequency	1	1.000
Second harmonic	½	0.500
Third harmonic	⅓	0.333
Fourth harmonic	¼	0.250
Fifth harmonic	⅕	0.200
Sixth harmonic	⅙	0.167
Seventh harmonic	⅐	0.143
Eighth harmonic	⅛	0.125

Ninth harmonic	$\frac{1}{9}$	0.111
Tenth harmonic	$\frac{1}{10}$	0.100
Eleventh harmonic	$\frac{1}{11}$	0.091
Twelfth harmonic	$\frac{1}{12}$	0.083

and so forth.

The relative amplitude of the harmonics is as important in defining the wave-shape as the harmonics that are or aren't included. For example, a triangle wave, shown in Fig. 3-6, includes the fundamental frequency and all odd harmonics, but no even harmonics—just like a square wave, but the harmonic amplitudes are much weaker. The amplitude of each harmonic in the square wave is equal to the amplitude of the fundamental frequency divided by the harmonic number:

$$A_h = \frac{A_f}{H}$$

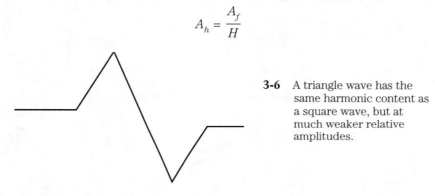

3-6   A triangle wave has the same harmonic content as a square wave, but at much weaker relative amplitudes.

But the amplitude of each harmonic in the triangle wave is equal to the amplitude of the fundamental frequency divided by the square of the harmonic number:

$$A_h = \frac{A_f}{H^2}$$

or:

$$A_h = \frac{A_f}{(H \times H)}$$

So we can compare a square wave and a triangle wave like this:

Fundamental frequency	1	1
Third harmonic	$\frac{1}{3}$	$\frac{1}{9}$
Fifth harmonic	$\frac{1}{5}$	$\frac{1}{25}$
Seventh harmonic	$\frac{1}{7}$	$\frac{1}{49}$
Ninth harmonic	$\frac{1}{9}$	$\frac{1}{81}$
Eleventh harmonic	$\frac{1}{11}$	$\frac{1}{121}$

and so forth.

The upper harmonics are much weaker in a triangle wave than in a square wave.

Because there are more similarities than differences in the circuit design, we will discuss rectangle-wave generators, square-wave generators, and pulse-wave generators together in the remainder of this chapter. If a given circuit is particularly suited for (or restricted to) square waves or pulse waves, this will be noted. Similarly, some rectangle-wave generators are not capable of generating true square waves. This limitation will also be noted, where relevant.

Transistor rectangle-wave generator circuits

To generate a rectangle wave, there must be some method of switching quickly between the two output levels (HIGH and LOW). A transistor is primarily a current amplifier, but it can function as a high-speed electronic switch. Therefore, it is suitable for rectangle-wave generation.

A very simple, but effective rectangle-wave generator circuit is shown in Fig. 3-7. Besides being very simple, this circuit is quite flexible in its component requirements. Nothing is too terribly critical here. Almost any general-purpose npn transistors can be used in this circuit, but both should be of the same type number, so that they have essentially the same operating characteristics. It doesn't matter so much what the specifics of those operating characteristics are, as long as they are reasonably balanced and matched within the circuit.

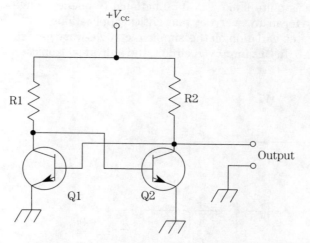

3-7 A very simple, but effective rectangle-wave generator circuit.

However, no two transistors are absolutely identical, even when they are of the same type number. This circuit is designed to take advantage of that fact. When power is first applied to this circuit, one of the two transistors will start to conduct slightly faster than its partner. It doesn't matter which one it is, but for the sake of our discussion, we will assume that Q1 conducts a little more heavily than Q2. The difference is very slight, but there will always be some difference.

At this point Q2 is cut off, because Q1's collector voltage is applied to the base of Q2. But, as Q1 continues to conduct more heavily, its collector voltage starts to drop. At some point, the base of Q2 is at a voltage that allows this transistor to turn on and start conducting. When Q2's conduction begins, the collector voltage of Q2 goes very high. This voltage is fed to the base of Q1, cutting the first transistor off.

Now, everything described in the previous paragraph is repeated, except that Q1 and Q2 have reversed roles. The two transistors switch back and forth in this fashion, turning each other on and off, for as long as power is applied to the circuit. If you tap off a signal from one of the transistor's collectors, you will get a rectangle wave. It doesn't matter which transistor you use for the output. In fact, you can use both, if that suits your application. The two collector signals are 180 degrees out-of-phase with each other. That is, when one is HIGH, the other is LOW, and vice versa.

The speed of the switching action (and therefore, the output signal frequency) is determined by the two collector resistors, which happen to be the only other components in this super-simple circuit. If these two resistors have equal values, each transistor will be on for an equal length of time, equal to one half the total cycle time. This means that the output will be a square wave. Rectangle waves with other duty cycles can easily be achieved, simply by using unequal resistor values for the two transistors. The ratio of the resistor values will determine the duty cycle of the output rectangle wave. For reliable performance, the resistance values should each be kept between 100 Ω and about 2.2 kΩ (2,200 ohms).

Obviously, being made up of just four components (two transistors and two resistors), this is about the simplest rectangle-wave generator circuit you are ever likely to come across. Of course, it is not terribly precise, either in terms of output frequency or waveshape, and it is quite limited in its functional frequency range. However it is perfectly adequate for many everyday non-critical applications.

A somewhat more sophisticated variation on this simple rectangle-wave generator circuit is illustrated in Fig. 3-8. In this improved circuit, the individual transistor

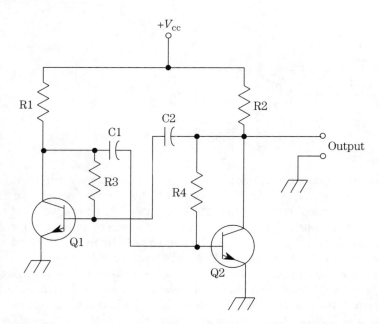

3-8   A somewhat more sophisticated variation on the simple rectangle-wave generator circuit of Fig. 3-7.

"on" times (and therefore the signal frequency and the duty cycle) is determined by the charging rates of the capacitors, along with the circuit resistances. Once again, using equal component values in both halves of the circuit will result in a square wave at the output.

The circuit shown in Fig. 3-9 is even more sophisticated and precise, at the inevitable expense of an increased component count. This circuit can be used reliably over a very wide range of output signal frequencies. Reasonably clean rectangle waves from about 0.5 Hz up to about 60 kHz (60,000 Hz) can be generated by this circuit.

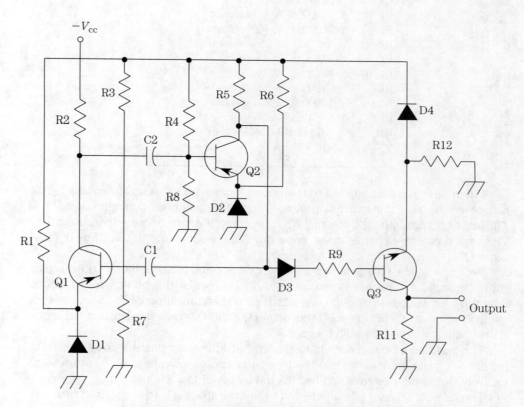

3-9 Here is an improved rectangle-wave generator circuit.

A standardized parts list for this circuit is given in Table 3-1. These component values are suitable for most practical applications. You will notice that three component values are omitted from this standardized parts list—capacitors C1, C2, and resistor R11. The reason for this omission is that these component values are selected for the specific desired output signal. The procedure for calculating these component values will be covered shortly. But first, let's take a moment to examine how this circuit functions.

Table 3-1.
Suggested parts list for the improved
rectangle-wave generator circuit of Fig. 3-9

Q1, Q2	pnp transistor (2N1303, or similar)
Q3	npn transistor, (2N1302, or similar)
D1, D2	Diode (1N463, or similar)
D3	Diode (1N95, or similar)
D4	Diode (1N645, or similar)
C1	See text
C2	See text
R1, R6	12-kΩ, ¼-W, 5% resistor
R2	360-Ω, 1-W, 10% resistor
R3, R4	3.6-kΩ, ¼-W, 5% resistor
R5	910 ¼-W, 5% resistor
R7, R8	620-Ω, ¼-W, 5% resistor
R9	510-Ω, ¼-W, 5% resistor
R10	100-Ω, ¼-W, 5% resistor
R11	See text
R12	5.6-kΩ, ¼-W, 5% resistor
V_{cc}	–12 V

When power is first applied to the circuit, transistor Q1 starts to conduct. The collector voltage of this transistor rises, charging capacitor C2 through resistor R4. This cuts off transistor Q2, causing its collector voltage to become more negative. This puts a negative charge across capacitor C1, speeding up the "on" time of transistor Q3.

Before long, the charge across capacitor C2 is equal to about 63% of the circuit's supply voltage. A capacitor is considered fully charged at this point. This capacitor now starts to discharge through Q1, which exhibits a low impedance. Now, the second half of the cycle begins, with transistors Q1 and Q2 trading roles, and capacitor C1 becoming charged through resistor R3.

The addition of transistor Q3 to this circuit limits any possible loading effects from the output circuit or device. The frequency and/or waveshape of the output signal from the simpler circuits can be affected by certain load conditions.

The value of resistor R11 is selected to suit the specific load being used with this circuit. The formula is:

$$R_{11} = \frac{V_o}{I_c 3 \times I_L}$$

where V_o is the desired output voltage, $I_c 3$ is the maximum collector current of transistor Q3 (from the manufacturer's specification sheet for the device), and I_L is the anticipated load current. If a 12-volt supply voltage is used with this circuit, V_o should be no more than I_L volts.

The output signal frequency is determined by the values of capacitors C1 and C2. The formula for finding the required value for C1 is:

$$C_1 = \frac{(I_c 1 \times T_L)}{(0.63 \times V_{cc})}$$

where $I_c 1$ is the collector current of transistor Q1, T_L is the desired LOW time per cycle (in seconds), and V_{cc} is the circuit's supply voltage.

The equation for capacitor C2 is similar:

$$C_2 = \frac{(I_c 2 \times T_H)}{(0.63 \times V_{cc})}$$

In this case, $I_c 2$ is the collector current of transistor Q2, T_H is the desired HIGH time per cycle (in seconds), and V_{cc} is the circuit's supply voltage.

Clearly, the total cycle time is simply the sum of the LOW time (determined by capacitor C1) and the HIGH time (determined by capacitor C2):

$$T_t = T_L + T_H$$

The output signal frequency is simply the reciprocal of the total cycle time:

$$F = \frac{1}{T_t}$$

$$= \frac{1}{(T_L + T_H)}$$

$$= \frac{1}{\left\{\left[\frac{(0.63 V_{cc} C_1)}{I_c 1}\right] + \left[\frac{(0.63 V_{cc} C_2)}{I_c 2}\right]\right\}}$$

We can simplify this equation considerably by standardizing some of the values. In most cases, the same transistor type number should be used for Q1 and Q2. This means their collector currents should be essentially equal. A typical value is 30 mA (0.03 ampere). A good standardized supply voltage for this circuit is 12 volts. Now, we can plug these standardized values into the equations and see what we are left with:

$$C_1 = \frac{(I_c 1 \times T_L)}{(0.63 \times V_{cc})}$$

$$= \frac{(0.03 \times T_L)}{(0.63 \times 12)}$$

$$= \frac{0.03 T_L}{7.56}$$

$$C_2 = \frac{(I_c 2 \times T_H)}{(0.63 \times V_{cc})}$$

$$= \frac{(0.03 \times T_H)}{(0.63 \times 12)}$$

$$= \frac{0.03 T_H}{7.56}$$

$$= \frac{1}{\left\{ \left[\frac{(0.63V_{cc}C_1)}{I_c1} \right] + \left[\frac{(0.63V_{cc}C_2)}{I_c2} \right] \right\}}$$

$$= \frac{1}{\left\{ \left[\frac{(0.63 \times 12 \times C_1)}{0.03} \right] + \left[\frac{(0.63 \times 12 \times C_2)}{0.03} \right] \right\}}$$

$$= \frac{1}{\left[\left(\frac{7.56C_1}{0.03} \right) \right] + \left[\left(\frac{7.56C_2}{0.03} \right) \right]}$$

These standardized equations are obviously a lot easier to work with. Let's try a couple very quick examples. Let's say we want to design this circuit to put out a square wave (duty cycle = 1:2) with an output signal frequency of 3.9 kHz (3,900 Hz). The total cycle time is:

$$T_t = \frac{1}{F}$$

$$= \frac{1}{3900}$$

$$= 0.0002564 \text{ second}$$

Because this will be a square wave, the LOW time and the HIGH time will each equal one half the total cycle time:

$$T_L = T_H = \frac{T_t}{2}$$

$$= \frac{0.0002564}{2}$$

$$= 0.0001282 \text{ second}$$

Now, we can find the necessary capacitor values:

$$C_1 = \frac{0.03T_L}{7.56}$$

$$= \frac{(0.03 \times 0.0001282)}{7.56}$$

$$= \frac{0.000003846}{7.56}$$

$$= 0.0000005 \text{ farad}$$

$$= 0.5 \text{ μF}$$

The nearest standard capacitor value, 0.47 μF, should be close enough.

We want the circuit to generate a square wave, with equal LOW and HIGH times, so capacitor C2 must obviously have the same value as C1. This will always be true for square waves of any frequency.

Now, let's consider a second practical example of this circuit. Let's assume C1 has a value of 0.22 µF and C2 has a value of 0.068 µF. The duty cycle ratio can be derived from the capacitor value ratio. The total capacitance is:

$$C_t = C_1 + C_2$$
$$= 0.22 + 0.068$$
$$= 0.288$$

The HIGH time is set by capacitor C2, so the duty cycle ratio will be the ratio between C_2 and C_t:

$$dc = C_2{:}C_t$$
$$= 0.068{:}0.288$$
$$= \left(\frac{0.068}{0.068}\right){:}\left(\frac{0.288}{0.068}\right)$$
$$= 1{:}4.24$$
$$= 1{:}4$$

The output signal frequency for this particular circuit works out to:

$$F = \frac{1}{\left[\left(\dfrac{7.56C_1}{0.03}\right)\right] + \left[\left(\dfrac{7.56C_2}{0.03}\right)\right]}$$

$$= \frac{1}{\left[\left[\dfrac{(7.56 \times 0.00000022)}{0.03}\right] + \left[\dfrac{(7.56 \times 0.000000068)}{0.03}\right]\right]}$$

$$= \frac{1}{\left[\left[\dfrac{(0.0000016)}{0.03}\right] + \left[\dfrac{(0.00000051)}{0.03}\right]\right]}$$

$$= \frac{1}{(0.0000533 + 0.0000166)}$$

$$= \frac{1}{0.0000699}$$

$$= 14{,}306 \text{ Hz}$$

Suppose your application requires fine-tuning the output frequency, or you need a variable frequency output. In such cases, replace resistors R3 and R4 with potentiometers or trimpots, with a limiting resistor in series with each.

A good choice for fine-tuning/calibration is a 2.2-kΩ (2,200 ohms) resistor in series with a 2.5-kΩ (2,500 ohms) trimpot for each R3 and R4. Calculate the nominal values for capacitors C1 and C2 in the usual manner. Then construct (or breadboard) the circuit and adjust the trimpots while monitoring the output signal's frequency with a frequency counter or an oscilloscope. For a variable-output frequency circuit, use a 1-kΩ (1,000 ohms) resistor in series with a 5-kΩ (5,000 ohms) or 10-kΩ (10,000 ohms) potentiometer for each R3 and R4.

There is one major disadvantage to this method of altering the frequency. The two controls will interact, affecting both the output signal frequency and the duty cycle. The fine-tuning procedure requires considerable patience in most cases. This type of circuit is really only suited for fixed-frequency applications.

Op amp square-wave generators

As with so many other applications in modern electronics, ICs can greatly simplify the design of rectangle-wave and square-wave generators. In this section, we will use op amps (operational amplifiers) to generate square waves. It is usually easiest to set up an op amp circuit to generate a symmetrical waveform, so the 1:2 duty cycle of the square wave comes naturally. In the next section, we will look at some tricks for getting an op amp to generate rectangle waves with other duty cycles.

Unless your application calls for extremely high frequencies that go well beyond the audible range, almost any general-purpose op amp device should work well in a square-wave generator circuit. In most applications, there would be little practical gain from using a more expensive high-grade op amp chip, although certainly no harm would be done, beyond the added cost. In any square-wave or rectangle-wave generator circuit, the most important specification for the op amp is probably the slew rate. It determines how rapidly the device can switch its output between the LOW and HIGH states. In other words, this determines how straight up and down the sides of the waveform (the transitions) will be.

One of the simplest square-wave generator circuits I've ever come across is shown in Fig. 3-10. This is about as simple as we can reasonably expect such a circuit to get. Besides the op amp itself, only three resistors and a single capacitor are needed to complete the circuit.

The most important factors in determining the output signal frequency of this circuit are the values of resistor R1 and capacitor C1, but the feedback network (R3 and R2) also affects the output frequency. In designing a circuit of this type, we must remember that all of the external passive components in this circuit have some de-

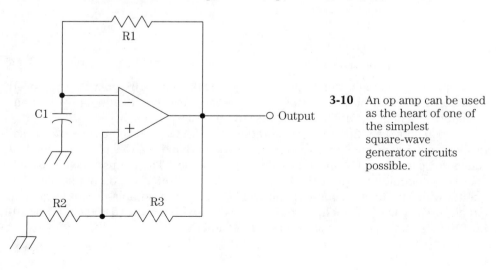

3-10 An op amp can be used as the heart of one of the simplest square-wave generator circuits possible.

gree of control over the output signal frequency. The exact equations are rather complex and difficult. Fortunately, we can simplify matters considerably if we limit ourselves to one of two basic R3:R2 ratios, or amplifier gains.

For example if resistors R2 and R3 are given identical values, their ratio will be 1:1. In this case, the formula for the output signal frequency can be simplified down to:

$$F = \frac{0.5}{(R_1 C_1)}$$

Alternately, the value of R3 can be made ten times greater than the value of R2, so the ratio is 10:1. For instance, if R2 is a 2.2-kΩ (2,200 ohms) resistor, R3 would be a 22-kΩ (22,000 ohms) resistor. With a feedback ratio of 10:1, the output signal frequency will be equal to:

$$F = \frac{5}{(R_1 C_1)}$$

Unless there is a very strong (and unusual) for doing otherwise, it is best to limit this circuit to one of these two feedback ratios—1:1, or 10:1.

As a typical example, let's design a square-wave generator circuit with an output signal frequency of 1,500 Hz (1.5 kHz), and a feedback ratio of 10. The first step in the circuit design process is to select an arbitrary value for resistor R2, then find an appropriate value for resistor R3:

$$R_3 = 10 R_2$$

Lets use a 33-kΩ (33,000 ohms) resistor for R2. This means the value of R3 will be:

$$R_3 = 10 \times 33000$$
$$= 330,000 \text{ ohms}$$
$$= 330 \text{ k}\Omega$$

Notice that if R2 is a standard resistor value, R3 will always be a standard resistor value, too.

Because the feedback ratio in this example is 10, we know the correct frequency equation for designing our circuit is:

$$F = \frac{5}{(R_1 C_1)}$$

We can select a likely value for capacitor C1, then algebraically rearrange the frequency equation to solve for the value of resistor R1:

$$R_1 = \frac{5}{(F C_1)}$$

In our example, we will use a 0.22-μF (0.00000022 μF) capacitor for C1. This means that the value of resistor R1 should be:

$$R_1 = \frac{5}{(1500 \times 0.00000022)}$$
$$= \frac{5}{0.00033}$$
$$= 15,152 \text{ ohms}$$

A standard 15-kΩ resistor should be close enough for most practical purposes. A standard 5%-tolerance resistor can be off by as much as 750 ohms, and still be considered within its acceptable range.

In some cases, you might come up with an awkward and impractical value for R1. If this happens, just try a new value for C1, and work through the equation again.

If your application calls for a variable frequency square wave signal, this circuit can easily be adapted to suit your needs. It's simply a matter of using a variable resistance element (such as a potentiometer) for R1. A variable frequency version of this simple op amp square-wave generator circuit is shown in Fig. 3-11. Not much has been changed here. The original R1 resistor has been replaced by R1a and R1b. R1a is a potentiometer, and R1b is a series resistor to limit the lowest possible resistance setting.

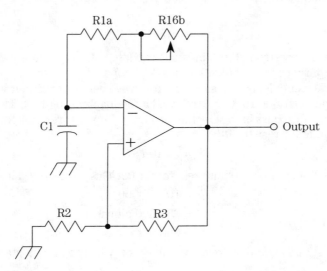

3-11 This is a variable frequency version of the simple op amp square-wave generator circuit of Fig. 3-10.

Let's assume we are working with this circuit using the following component values:

R1a = 100-kΩ potentiometer (100,000 ohms)
R1b = 10-kΩ resistor
R2 = 47-kΩ resistor
R3 = 47-kΩ resistor
C1 = 0.01 μF capacitor

Notice that resistors R2 and R3 have identical values, so the feedback ratio is 1, so the frequency equation we need to use here is:

$$F = \frac{0.5}{(R_1 C_1)}$$

The value of R1 is the series combination of R1a and R1b. Resistances in series always add:

$$R_1 = R_{1a} + R_{1b}$$

The value of R1b is constant, but R1a's resistance varies as the potentiometer's shaft is rotated. At its lowest setting, the resistance value will be close to zero. Actually, it will probably be a couple hundred ohms, but this makes little practical difference. We might as well just assume it is zero. This means the value of R1 is simply equal to the value of R1b:

$$R_1 = 0 + 10000$$
$$= 10,000 \text{ ohms}$$

So the output signal frequency for this setting of potentiometer R1a works out to:

$$F = \frac{0.5}{(10000 \times 0.00000001)}$$
$$= \frac{0.5}{0.0001}$$
$$= 5,000 \text{ Hz}$$
$$= 5 \text{ kHz}$$

This is the highest frequency that this circuit can generate.

At the opposite extreme setting, the potentiometer (R1a) will have its full rated resistance value, or 100 kΩ, so the total value of R1 is:

$$R_1 = 100000 + 10000$$
$$= 110,000 \text{ ohms}$$

At this setting of potentiometer R1a, the output frequency is equal to:

$$F = \frac{0.5}{(110000 \times 0.00000001)}$$
$$= \frac{0.5}{0.0011}$$
$$= 450 \text{ Hz}$$

So, by using this set of component values, our simple square-wave generator circuit can cover a little more than a 4.5-kHz range—from 450 Hz up to 5 kHz (5,000 Hz). This is a lot simpler than trying to get a variable frequency from any of the transistor-based rectangle-wave generator circuits discussed in the preceding section.

For an even wider range of output frequencies, you can use a rotary switch (or something similar) to select between several different capacitors with varied values. Whichever capacitor is selected by the switch acts like "C1" as far as the frequency equation is concerned.

Op amp rectangle-wave generators

This basic op-amp circuit always produces 1:2 duty cycle square waves because the same feedback path is used for both the negative and positive half-cycles, so their timing periods are inevitably equal. If we want to generate rectangle waves with some other duty cycle, we will need to set up different feedback paths for the half-cycles. A very easy way to accomplish this is illustrated in the modified circuit of Fig. 3-12.

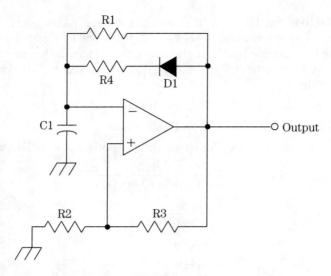

3-12 This circuit demonstrates a very easy way to generate unequal duty-cycle rectangle waves with an op amp.

On the negative half-cycles, diode D1 is reverse-biased and blocks the flow of
current through resistor R4. In effect, this resistor has been deleted from the circuit.
The time constant for the negative half-cycle is determined by the values of resistor
R1 and capacitor C1, as in the original version of the circuit:

$$T_L = C_1 \times R_1$$

On the positive half-cycles, however, the diode is forward-biased, so it conducts,
and resistor R4 is effectively placed in parallel with resistor R1. (In most practical applications, we can reasonably ignore the internal resistance of the diode itself.) This
means that the time constant for the HIGH portion of the cycle is determined by the
value of capacitor C1, and both resistors R1 and R4:

$$T_H = C_1 \times \left[\frac{(R_1 \times R_4)}{(R_1 + R_4)} \right]$$

Because the time constant is different for the positive and negative half-cycles,
the duty cycle will be something other than 1:2. The circuit will generate a non-square
rectangle wave. The total time constant of the circuit is simply the sum of the LOW
and HIGH time constants, of course, so the frequency equation works out to either:

$$F = \frac{0.5}{(T_L + T_H)}$$

or:

$$= \frac{0.5}{\left\{ (C_1 R_1) + C_1 \left[\frac{(R_1 \times R_4)}{(R_1 + R_4)} \right] \right\}}$$

$$F = \frac{0.5}{(T_L + T_H)}$$

$$= \frac{5}{\left\{ (C_1 R_1) + C_1 \left[\frac{(R_1 \times R_4)}{(R_1 + R_4)} \right] \right\}}$$

depending on the feedback ratio set up by the values of resistors R2 and R3, as in the square-wave version described in the last section. Let's assume R2 and R3 have identical values, giving a feedback ratio of 1:1. In this case, the correct frequency equation is:

$$F = \frac{5}{\left\{ (C_1 R_1) + C_1 \left[\frac{(R_1 \times R_4)}{(R_1 + R_4)} \right] \right\}}$$

Let's further assume the following component values in the circuit:

C1 = 0.047 µF   (0.000000047 farad)
R1 = 22 kΩ   (22,000 ohms)
R2 = 10 kΩ   (10,000 ohms)
R3 = 10 kΩ   (10,000 ohms)
R4 = 39 kΩ   (39,000 ohms)

Solving for the LOW time constant we get a value of:

$$T_1 = C_1 \times R_1$$

$$= 0.000000047 \times 22000$$

$$= 0.001034 \text{ second}$$

$$= 1.034 \text{ mS}$$

Next, we solve for the HIGH time constant, which is a little more complex, because resistor R4 is added in parallel with resistor R1:

$$T_H = C_1 \times \left[\frac{(R_1 \times R_4)}{(R_1 + R_4)} \right]$$

$$= 0.000000047 \times \left[\frac{(22000 \times 39000)}{(22000 + 39000)} \right]$$

$$= 0.000000047 \times \left(\frac{858000000}{61000} \right)$$

$$= 0.000000047 \times 14066$$

$$= 0.000661 \text{ second}$$

$$= 0.661 \text{ mS}$$

Finally, we can put these values together to solve for the output signal frequency:

$$F = \frac{0.5}{(T_L + T_H)}$$

$$= \frac{0.5}{(0.0010134 + 0.000661)}$$

$$= \frac{0.5}{0.001695}$$

$$= 295 \text{ Hz}$$

Either or both of the frequency-determining resistors can be made variable. Unfortunately, the controls will interact, and changing the frequency will alter the duty cycle, and vice versa.

An improved version of this same circuit is shown in Fig. 3-13. Here we are using two diodes, so only one resistor is in the active feedback path on either half-cycle. This means that the two time constants can be 100% independent of each other. The calculations are also simpler, because there no parallel resistances to worry about:

$$T_L = C_1 \times R_1$$
$$T_H = C_1 \times R_4$$

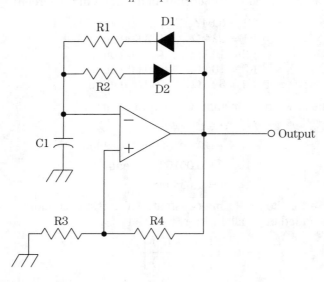

3-13   This improved version of the rectangle-wave generator circuit of Fig. 3-12 permits independent control of the length of each half-cycle.

$$\text{If } R_3 = R_2$$

$$\text{Then } F = \frac{0.5}{(T_L + T_H)}$$

$$= \frac{0.5}{[(C_1 \times R_1) + (C_1 \times R_4)]}$$

$$= \frac{0.5}{[C_1(R_1 + R_4)]}$$

or:

$$\text{If } R_3 = 10R_2$$

$$\text{Then } F = \frac{5}{(T_L + T_H)}$$

$$= \frac{5}{[(C_1 \times R_1) + (C_1 \times R_4)]}$$

$$= \frac{5}{[C_1(R_1 + R_4)]}$$

555 astable multivibrators

A rectangle-wave generator is sometimes called an *astable multivibrator*. A multivibrator is a general type of circuit with just two possible output states—HIGH and LOW. There are three types of multivibrators:

Monostable multivibrator	One stable output state
Bistable multivibrator	Two stable output states
Astable multivibrator	No stable output states

In this book we are only concerned with astable multivibrators, in which neither output is stable. The output continuously switches back and forth between its two possible states at a rate determined by timing component values within the multivibrator circuit. In other words, the astable multivibrator generates a rectangle wave output signal.

The 555 timer is one of the most popular ICs now on the market. It is designed primarily for monostable multivibrator applications, but it works equally well with astable multivibrator circuits. This chip is very inexpensive, widely available, and extremely easy to work with. It can be operated over a fairly wide frequency range, including well over the entire audible frequency spectrum. The supply voltage for a 555 timer circuit can be anything from +4.5 volts up to +15 volts.

The pin-out diagram of the 555 timer IC is shown in Fig. 3-14. This device is also available in a dual version called the 556, shown in Fig. 3-15. The 556 dual-timer chip contains two separate 555-type timer circuits in a single housing. To substitute a 556 for two 555s, or vice versa, all you need to do is correct the pin numbers accordingly. The two chips are electrically identical, and no external circuitry changes are ever necessary to substitute one for the other.

The basic 555 astable multivibrator circuit is shown in Fig. 3-16. As you can see, this is a very simple circuit, and it is complete and functional as shown here. Other than the 555 IC itself, only four external components are needed—two resistors and two capacitors.

As a matter of fact, capacitor C2 is optional in this circuit. It is included solely to improve the stability of the circuit, which will rarely be a problem in any case. In most cases, this capacitor would not be missed at all if it was omitted. Still, it is a good idea to always include capacitor C2 in all 555 timer circuits that do not use the voltage control input (pin #5). It is cheap insurance against possibly frustrating

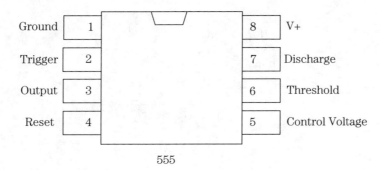

3-14 The 555 timer is one of the most popular ICs now on the market.

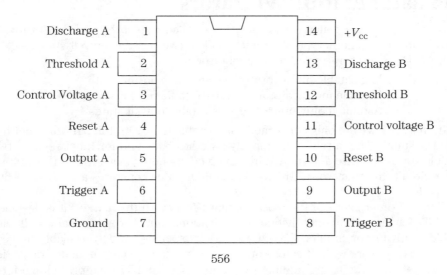

3-15 The 556 dual-timer chip contains two separate 555-type timer circuits in a single housing.

problems that could show up if there is any instability in the circuit. This could cause very erratic operation. If you use a cheap ceramic disc capacitor for C2 (and there's no good reason to use anything better in this application), it only adds a dime or two to the overall circuit cost. It's hardly worth scrimping on.

The exact value of this stability capacitor is far from critical. In most cases, anything from about 0.001 µF and 0.047 µF can be used equally well. This capacitance is usually standardized as 0.01 µF, but this is not important. The other three external components in this circuit (resistors R1 and R2, along with capacitor C1) determine the output frequency and the duty cycle.

Notice also that the 555's trigger input (pin #2) is shorted to the threshold input (pin #6). This connection forces the timer to operate in its astable mode.

When power is first applied to this circuit, the voltage across timer C1 is low. As a result of this low voltage, the timer is triggered (through pin #2). The output goes to its HIGH state, and the IC's internal discharge transistor (inside the chip at pin #7)

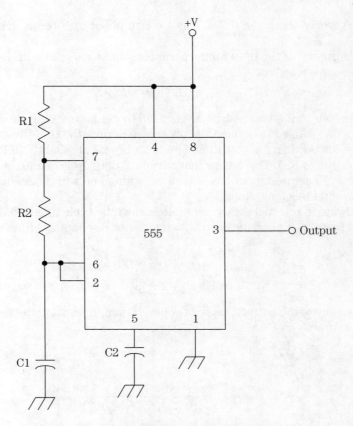

3-16 This is the basic 555 astable multivibrator circuit.

is turned off. A complete current path through C1, R1, and R2 is formed, charging the capacitor.

When the charge on capacitor C1 exceeds two-thirds (67%) of the supply voltage, the timer's upper threshold is reached. This voltage on pin #6 forces the output back to its LOW state.

The timing capacitor now starts to discharge through resistor R2, but not through R1 because the internal transistor (at pin #7) is now turned on. When the voltage across C1 drops below one third (33%) of the supply voltage, the timer is automatically retriggered, and a new cycle begins.

The exact supply voltage used is not critical. The timing is derived from internally determined ratios within the 555 timer chip itself. Therefore, the actual supply voltage does not noticeably affect the timing cycles. Although, if the supply voltage should happen to change value for any reason, the timing is likely to be affected until the circuit gets a change to restabilize to its new supply voltage. This is not recommended.

The LOW timing period is determined by resistor R2 and capacitor C1 according to this formula:

$$T_L = 0.693 R_2 C_1$$

Don't worry about the 0.693. It is a constant set by internal values within the 555 IC.

The formula for the HIGH timing period is similar, except in this case, both resistor values have an effect:

$$T_H = 0.693(R_1 + R_2)C_1$$

Since the sum of the two resistances is used to define the HIGH timing period, and only R2 alone is used to define the LOW timing period, the HIGH timing period must always be longer than the LOW timing period, even if only slightly (if R1 is very small in comparison to R2). This means that a true 1:2 duty cycle cannot be achieved with this circuit. Some designers have come up with some very ingenious methods for getting around this limitation.

As always in rectangle-wave generators, the total cycle time is simply the sum of the HIGH (capacitor charging) and LOW (capacitor discharging) time periods:

$$T_t = T_H + T_L$$
$$= 0.693(R_1 + R_2)C_1 + 0.693R_2C_1$$
$$= 0.693(R_1 + 2R_2)C_1$$

The frequency of the output signal, of course, is simply equal to the reciprocal of the total cycle time:

$$F = \frac{1}{T_t}$$

$$= \frac{1}{(0.693(R_1 + 2R_2)C_1)}$$

$$= \frac{1.44}{[(R_1 + 2R_2)C_1]}$$

For proper operation, these component values should be held between specific limits. The combined series value (sum) of resistors should be no less than 10 kΩ (10,000 ohms), and no more than 14 MΩ (14,000 ohms). Neither resistor should be less than 1 kΩ (1,000 ohms). The timing capacitor (C1) should have a value between about 100 pF (0.0000000001 farad) and approximately 1,000 μF (0.001 farad).

For our first practical example, we will be using the minimum standard values throughout the circuit:

R1 = 9.1 kΩ   (9,100 ohms)
R2 = 1 kΩ   (1,000 ohms)
C1 = 100 pF   (0.0000000001 farad)

In this case, the output frequency works out to:

$$F = \frac{1.44}{[(9100 + 2 \times 1000) \times 0.0000000001]}$$

$$= \frac{1.44}{[(9100 + 2000) \times 0.0000000001]}$$

$$= \frac{1.44}{(11100 \times 0.0000000001)}$$

$$= \frac{1.44}{0.00000000111}$$

$$= 129{,}729{,}729 \text{ Hz}$$

(Note, the 555 timer circuit cannot be expected to operate reliably or stably at frequencies above about 1 MHz (1,000,000 Hz)—which is usually (though not always) the maximum practical output frequency.

For our second example, we will go to the opposite extreme, and use the maximum acceptable component values:

$$R1 = 3.9 \text{ M}\Omega \qquad (3{,}900{,}000 \text{ ohms})$$
$$R2 = 10 \text{ M}\Omega \qquad (10{,}000{,}000 \text{ ohms})$$
$$C1 = 1000 \text{ }\mu\text{F} \qquad (0.001 \text{ farad})$$

In this case, the output frequency works out to:

$$F = \frac{1.44}{[(3{,}900{,}000 + 2 \times 10{,}000{,}000) \times 0.001]}$$

$$= \frac{1.44}{[(3{,}900{,}000 + 20{,}000{,}000) \times 0.001]}$$

$$= \frac{1.44}{(23{,}900{,}000 \times 0.001)}$$

$$= \frac{1.44}{23{,}900}$$

$$= 0.00006 \text{ Hz}$$

Or about one cycle every 16,667 seconds or 4.6 hours.

As you can see, the 555 astable multivibrator circuit can theoretically cover a very impressive frequency range. In practical applications, however, I would strongly recommend keeping the output signal frequency between about 0.0001 Hz and 1 MHz (1,000,000 Hz)—still a remarkable frequency range compared to most common signal-generator circuits.

The 555 timer is well-suited for use in both analog and digital applications. It can normally use the same power supply as well. The 555 itself is an analog device, but a digital CMOS version also exists. This is the 7555. It is pin-for-pin compatible with the standard 555 IC, (the pin out diagram is exactly the same) and no changes in the external circuitry are required. All equations are the same for the 7555 as for the 555, but the 7555 is designed to meet CMOS circuit specifications.

The LM3909 LED flasher/oscillator

The LM3909 is a special oscillator chip that is incredibly easy to use. It was designed primarily for LED flasher applications, but it also has a fast-rate mode that permits its use as an audio oscillator. It's output signal will be in the form of square waves. The pin-out diagram for the LM3909 is shown in Fig. 3-17, and the basic LM3909 audio oscillator circuit appears as Fig. 3-18.

Notice the very low supply voltage. It can be increased up to +6.3 volts, but no higher, or the chip could be damaged.

The larger the capacitor valve, the lower the output signal frequency will be. Unfortunately, there is no convenient, simple frequency-determining formula for determining the output frequency of the LM3909. A very rough approximation can be found with this simple formula:

$$F = \frac{300}{C}$$

where F is the frequency in Hertz, and C is the capacitance in microfarads (not farads in this equation). Remember, the results of this equation are only approximate.

The resistor values also affect the output signal frequency in a lesser way. Potentiometer R1 permits fine-tuning of the output frequency to suit the desired application.

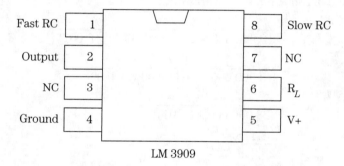

3-17 The LM3909 LED flasher/oscillator is very easy to use.

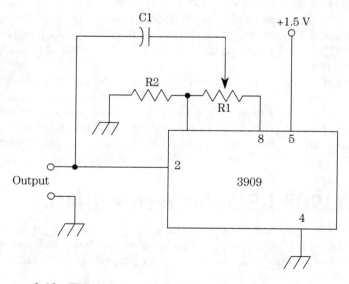

3-18 This is the basic LM3909 audio oscillator circuit.

Project #4—Variable-frequency/variable-duty cycle rectangle-wave generator

Most of the rectangle-wave generators described in this chapter permit at least some limited control over the output signal frequency. A few also permit the user to alter the duty cycle of the output waveform. Unfortunately, in the rectangle-wave generator circuits described so far, these two parameters interact. Whenever you change one, the other is affected. Obviously, this means adjusting the circuit to put out exactly the desired rectangle wave signal is difficult and rather time consuming.

In many applications, it is highly desirable to be able to independently set the output signal frequency and the duty cycle. That is the purpose of this project.

This project, which is illustrated in Fig. 3-19, utilizes both halves of a 556 dual-

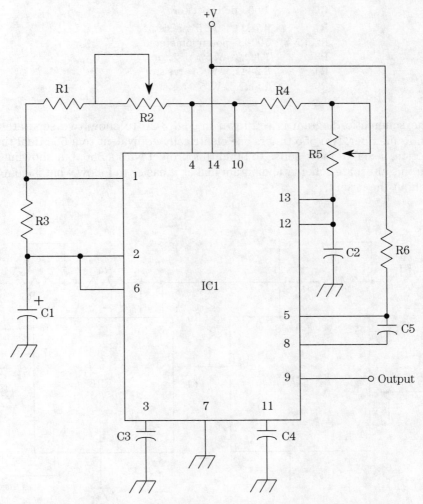

3-19 Project #4—Variable-frequency/variable duty-cycle rectangle-wave generator.

timer IC. In effect, one controls the output signal frequency and the other controls the duty cycle. A suitable parts list for this project is given in Table 3-2. You should feel free to experiment with any or all of the passive component values throughout the circuit, within the constraints described below. Nothing is terribly critical in this circuit.

Table 3-2.
Suggested parts list for Project #4—
variable-frequency/variable duty-cycle
rectangle-wave generator of Fig. 3-19

IC1	556 dual timer
C1	1-µF 35-V electrolytic capacitor
C2	0.1-µF capacitor
C3, C4	0.01-µF capacitor
C5	1,000-pF capacitor
R1	4.7-kΩ, ¼-W, 5% resistor
R2, R5	100-kΩ potentiometer
R3	1-kΩ, ¼-W, 5% resistor
R4	8.2-kΩ, ¼-W, 5% resistor
R6	10-kΩ, ¼-W, 5% resistor

The schematic diagram is re-drawn in Fig. 3-20 to show two separate 555 timers. Remember, two 555 timers are electrically equivalent to a 556 dual timer. In practice, it makes more sense to build this project with a 556, but showing two 555s in this alternate schematic diagram makes it easier to follow what is going on throughout the circuit.

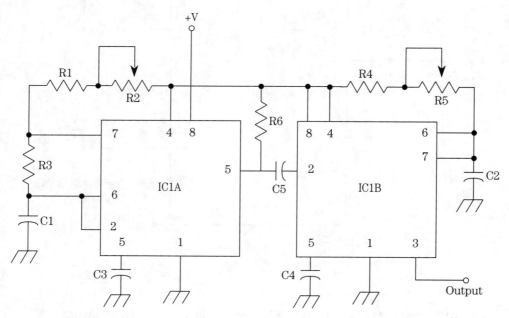

3-20   The circuit of Project #4 is redrawn here to show separate 555 timers.

The first timer stage (IC1A) is wired as a standard astable multivibrator. This is the same basic 555 rectangle-wave generator circuit we examined earlier in this chapter. This sub-circuit features a manually variable frequency through potentiometer R2. Series resistor R1 sets the minimum resistance so the potentiometer can't be accidentally set for too low a value. The value of resistor R3 is relatively small, so the duty cycle of the waveform generated by this sub-circuit is fairly narrow, even though the exact duty cycle varies as the signal frequency is changed. We don't have anything special yet.

The output signal from this astable multivibrator drives the trigger input of the second timer stage (IC1B), which is wired as a monostable multivibrator. A monostable multivibrator has just a single stable output state, which in this case is LOW. The output will normally be LOW all the time, until the multivibrator receives an external trigger pulse at its input. When triggered, the output will go HIGH for a specific time period determined by component values within the monostable multivibrator circuit—in this case resistors R4 and R5, and capacitor C2. The output time period will be constant, regardless of how long the input pulse lasts. In this application, the actual duty cycle of the signal from IC1A isn't very important. It doesn't matter that it changes whenever the frequency is adjusted (via potentiometer R2).

The timing period of the monostable multivibrator section (IC1B) in this circuit can be adjusted via potentiometer R5. Series resistor R4 prevents this potentiometer from being inadvertently set for too low a value. The output time period from this sub-circuit can be calculated with this simple equation:

$$T = 1.1RC_2$$

where R is the series combination of R_4 and R_5 in ohms, C_2 is in farads, and the time period T, is in seconds.

Assume for a moment that the value of potentiometer R5 is fixed. This means the timing period of the monostable multivibrator stage will always be the same, regardless of the length of the trigger pulses being generated by the astable multivibrator stage (IC1A). When the signal frequency is varied by means of potentiometer R2, the pulse width of the output signal will not change.

The actual pulse width (and therefore the duty cycle) of the output signal can be manually varied with potentiometer R5. Adjusting this control does not affect the signal frequency in any way.

Once per cycle, IC1A will trigger IC1B, so the circuit will put out one pulse with a timing period controlled by potentiometer R5. The frequency of these output pulses is controlled by potentiometer R2.

The only real restriction here is that the timing period of IC1B must always be less than the total cycle time of the signal generated by IC1A. Otherwise, the pulse of one cycle with overlap with the next cycle's pulse, and there will be no useful output from the circuit. The output will simply be HIGH all the time.

To avoid such problems, try not to adjust the circuit for very high frequencies and/or very long pulse lengths. If your project should happen to lock-up in this way, there should be no permanent harm done. Just re-adjust the setting of one or both of the potentiometers (R2 and R5) until you get a rectangle wave at the circuit's output. The component values suggested in the parts list were selected to minimize the likelihood of such lock-up problems.

Let's assume that each potentiometer is set to the midpoint of its range (50 kΩ). What will be the output signal frequency and the duty cycle generated by the circuit?

First, we solve for the frequency, using the standard 555 astable multivibrator frequency equation:

$$F = \frac{1.44}{[C_1(R_a + 2R_b)]}$$

In this circuit, R_a is the series combination of resistor R_1 and potentiometer R_2:

$$R_a = R_1 + R_2$$
$$= 4{,}700 + 50{,}000$$
$$= 54{,}700 \text{ ohms}$$

R_b is simply resistor R_3:

$$R_b = R_3$$
$$= 1 \text{ k}\Omega$$
$$= 1{,}000 \text{ ohms}$$

And C_1, of course, is the value of that capacitor in farads. According to the suggested parts list, this is a 1-μF (0.000001 farad) capacitor. In this circuit, this capacitor should be relatively large in order to keep the signal frequency appropriately low.

Plugging all of these component values into the frequency equation, we find the signal frequency generated by IC1A is equal to about:

$$F = \frac{1.44}{[0.000001(54{,}700 + 2 \times 1{,}000)]}$$

$$= \frac{1.44}{[0.000001(54{,}700 + 2{,}000)]}$$

$$= \frac{1.44}{(0.000001 \times 56{,}700)}$$

$$= \frac{1.44}{0.0567}$$

$$= 25 \text{ Hz}$$

The time period of each complete cycle is equal to the reciprocal of the signal frequency:

$$T_c = \frac{1}{F}$$

$$= \frac{1}{25}$$

$$= 0.04 \text{ second}$$

$$= 40 \text{ mS}$$

Meanwhile, the monostable multivibrator stage (IC1B) is using the following component values:

$$C_2 = 0.1 \ \mu F$$
$$= 0.0000001 \ \text{farad}$$
$$R = R_4 + R_5$$
$$= 8,200 + 50,000$$
$$= 58.200 \ \text{ohms}$$

Plugging this component values into the monostable timing formula, we get a timing period for the output HIGH pulse equal to about:

$$T_p = 1.1 \times 58,200 \times 0.0000001$$
$$= 0.0064 \ \text{second}$$
$$= 6.4 \ \text{mS}$$

We can compare the HIGH pulse time (T_p) to the total cycle time (T_c), to find the duty cycle:

$$dc = 6.4{:}40$$
$$= \left(\frac{6.4}{6.4}\right){:}\left(\frac{40}{6.4}\right)$$
$$= 1{:}6$$

Now, let's reduce the setting of potentiometer R5, without touching potentiometer R2. R2 remains unchanged, so the frequency (and total cycle time) are exactly the same as before. Let's say potentiometer R5 is adjusted so the total series value of R is now 12,000 ohms. Now the pulse length will be equal to:

$$T_p = 1.1 \times 12,000 \times 0.0000001$$
$$= 0.0013 \ \text{second}$$
$$= 1.3 \ \text{mS}$$

The total cycle time is still 40 milliseconds, so the duty cycle is now:

$$dc = 1.3{:}40$$
$$= \left(\frac{1.3}{1.3}\right){:}\left(\frac{40}{1.3}\right)$$
$$= 1{:}31$$

The output signal frequency is still 25 Hz, but the duty cycle has been changed from 1:6 to 1:31.

If we adjust potentiometer R2, but leave potentiometer R5 alone, the signal frequency will change, but not the length of the HIGH output pulse. Actually, this will affect the duty cycle somewhat, but it will still be more consistent than is usual in simple rectangle-wave generator circuits. And we can adjust the duty cycle independently of the frequency.

As a final example, let's leave the monostable multivibrator (IC1B) set up as in the last example, giving an output pulse of 1.3 mS, but we will now adjust potentiometer R2 so that Ra has an effective series value of 38 kΩ (38,000 ohms). This means the signal frequency is now equal to about:

$$F = \frac{1.44}{[0.0000001(38,000 + 2 \times 1,000)]}$$

$$= \frac{1.44}{[0.0000001(38,000 + 2,000)]}$$

$$= \frac{1.44}{(0.0000001 \times 40,000)}$$

$$= \frac{1.44}{0.004}$$

$$= 360 \text{ Hz}$$

The total cycle time is equal to:

$$T_c = \frac{1}{360}$$

$$= 0.0028 \text{ second}$$

$$= 2.8 \text{ mS}$$

The HIGH pulse time (T_p) is still 1.3 mS, so the duty cycle of the output signal now works out to approximately:

$$dc = 1.3{:}2.8$$

$$= \left(\frac{1.3}{1.3}\right){:}\left(\frac{2.8}{1.3}\right)$$

$$= 1{:}2$$

In summary, adjusting potentiometer R2 will change the signal frequency (and therefore the total cycle time), but not the pulse width. Or, adjusting potentiometer R5 will change the pulse width, but not the signal frequency (or total cycle time).

4
CHAPTER

Function generators

The signal-generator circuits we have looked at so far are designed to put out just one standard waveform. Sometimes, it is useful to have a circuit that can generate two or more different waveforms. Such a device is known as a *function generator*.

Some function generators can generate multiple waveforms simultaneously. They will all have the same frequency, and usually (though not always) the same phase. Other function generators are designed to emit just one of several waveforms at a time. The desired waveform is selected in such instruments with some type of manual or electronic switching mechanism.

Standard waveforms

Most commercial function generators produce two or more standard ac waveforms:

- Sine wave
- Rectangle wave (square wave or pulse wave)
- Triangle wave
- Sawtooth wave

Some circuits produce other, less common waveforms that often have no standardized names. We have already encountered the sine wave in chapter 2, and the various forms of the rectangle wave in chapter 3.

The *triangle wave* is sometimes known as the *delta wave*. Both names are derived from this waveshape's appearance when displayed on an oscilloscope, as illustrated in Fig. 4-1. Like the square wave, the triangle wave is comprised of a fundamental frequency and all odd harmonics, but no even harmonics. The difference is that the relative amplitude of the harmonic components are considerably weaker than in a square wave. In fact, the harmonic content of a triangle

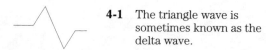

4-1 The triangle wave is sometimes known as the delta wave.

wave is so weak that this waveform is often used as a substitute for a sine wave (which consists of just the fundamental frequency, and theoretically has no harmonic content at all). The resemblance of a triangle wave to a sine wave is even closer if the signal is passed through a low-pass filter. In many practical function-generator circuits, it is much easier to generate a triangle wave than it is to generate a true sine wave.

Another popular standard waveform is the *sawtooth wave*. There are two types of sawtooth waves, *ascending* and *descending*. An ascending sawtooth wave (the more commonly used form) is shown in Fig. 4-2. You can see the visual relationship between this waveshape and the teeth of a saw. It is also called a *ramp wave*, again from the appearance. The instantaneous signal level starts at a specific minimum level (the negative peak) and builds smoothly and linearly up to a specific maximum level (the positive peak). As soon as the positive peak level is achieved, the signal drops instantly down to the negative peak, and the next cycle begins.

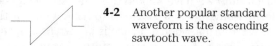

4-2 Another popular standard waveform is the ascending sawtooth wave.

This waveform is comprised of the fundamental frequency and all harmonics, both even and odd, all at relatively high amplitudes. Any desired harmonic or group of harmonics can easily be isolated with appropriate filter circuits. This process is known as *subtractive synthesis*.

A descending sawtooth wave is shown in Fig. 4-3. The difference here is the angle of the ramp. This signal begins its cycle at the positive peak, then drops smoothly and linearly down to the negative peak, then instantly jumps back up to the positive peak, and begins the next cycle.

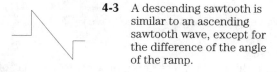

4-3 A descending sawtooth is similar to an ascending sawtooth wave, except for the difference of the angle of the ramp.

Like the ascending sawtooth wave, the descending sawtooth wave consists of the fundamental frequency and all harmonics, both even and odd, all at relatively high amplitudes. The phase relationships between the fundamental frequency and the various harmonics are somewhat different in the descending version of this waveform.

Most practical function generators are designed to produce two or more of the four standard ac waveforms. Other waveforms are sometimes encountered, such as the staircase waveforms illustrated in Fig. 4-4.

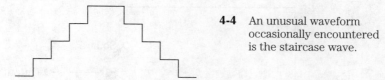

4-4 An unusual waveform occasionally encountered is the staircase wave.

Op amp function generators

In earlier chapters, we used op amps to generate sine waves and rectangle waves. This device is capable of generating any of the four standard waveforms discussed above (and many less common waveforms, as well).

The easiest way to generate a triangle wave in an op amp circuit is to use two op amps. The first one generates a square wave (as in the circuits of chapter 3), while the second op amp acts like a low-pass filter (or integrator circuit), adding increasing amounts of attenuation as the signal frequency is increased. A very gradual cut-off slope is used. Remember, a triangle wave and a square wave have the exact same harmonic make-up (the fundamental frequency and all odd harmonics, but no even harmonics) except that the relative amplitudes of the harmonics in the triangle wave are much weaker than in a square wave. The filter's cut-off frequency is selected to pass the fundamental frequency at full amplitude, and increasingly decrease the amplitude of the harmonics as they get further from the waveform's fundamental frequency.

A simple block diagram of this approach is shown in Fig. 4-5. We already have a square wave before we convert it into the triangle wave, and can use either waveform as an output signal. In other words, this is a two-waveform function generator. The two output waveforms are tapped off from different points within the circuit, so they can both be used simultaneously, provided excessive loads aren't placed on the function-generator circuit. Both output waveforms will always have the same frequency, and essentially the same phase.

In some applications, it might be preferable to have the option of switching a single output between the two available waveforms. This is easy enough to accomplish with a simple SPDT switch, as illustrated in Fig. 4-6.

A practical two-waveform function-generator circuit of this type is shown in Fig. 4-7. As you can see, this is a pretty simple circuit, requiring just two op amps, three

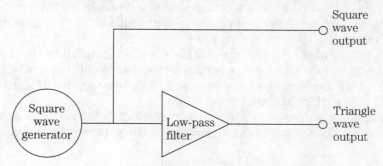

4-5 A simple function-generator circuit starts with a square wave, then filters it to produce a second, triangle wave output.

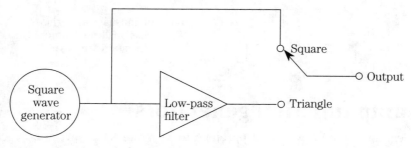

4-6   In some applications, it might be preferable to have the option of switching a single output between the two available waveforms.

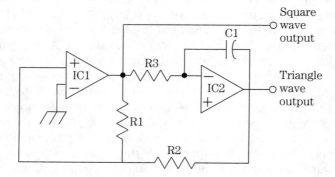

4-7   This is a practical two-waveform function-generator circuit.

resistors and a single capacitor. You can use two separate op amp ICs, or a dual op amp chip.

The output of IC1 is the square wave signal, while the output of IC2 is the triangle wave. The square wave signal swings back and forth between the op amp's positive and negative saturation voltages, which are typically just slightly below the actual supply voltages. If a single-ended power supply is used (this is not always possible with some op amp devices) the LOW signal level will be just slightly above ground potential (0 volts) and the HIGH signal level will be a little less than V+. When a dual-polarity supply voltage is used, the output signal will swing essentially between –V and +V.

The amplitude of the triangle wave output signal is controlled by the ratio of resistor values R_1 and R_2. These components also have some effect on the output frequency (for both the square waves and the triangle waves), and this makes the adjustment a little tricky in certain applications.

The output signal frequency is primarily controlled by the values of resistor R3 and capacitor C1. The complete output signal frequency formula for this circuit is:

$$F = \left[\frac{1}{(4R_3C_1)} \right] \times \left(\frac{R_1}{R_2} \right)$$

Let's consider a specific example. Our simple function-generator circuit will use the following component values:

$$C1 = 0.1\ \mu F \quad (0.0000001\ \text{farad})$$
$$R1 = 33\ k\Omega \quad (33,000\ \text{ohms})$$
$$R2 = 3.3\ k\Omega \quad (3,300\ \text{ohms})$$
$$R3 = 22\ k\Omega \quad (22,000\ \text{ohms})$$

In this case, the output signal frequency works out to a value of about:

$$F = \left[\frac{1}{(4 \times 22000 \times 0.0000001)} \right] \times \left(\frac{33000}{3300} \right)$$

$$= \left(\frac{1}{0.0088} \right) \times 10$$

$$= 113.64 \times 10$$
$$= 1,136.4\ \text{Hz}$$

The easiest and most effective way to manually vary the output signal frequency is to change the resistance of R3. Use a potentiometer in the place of the fixed resistor shown here. I recommend adding a small fixed resistor in series with the potentiometer. This prevents the possibility of setting the resistance value too low, and sets a lower limit for the output frequency range.

Let's return to our example, and keep everything the same as before, except we will change the value of R3 to 9.5 kΩ (9,500 ohms). This means that the output signal frequency will be changed to:

$$F = \left[\frac{1}{(4 \times 9500 \times 0.0000001)} \right] \times \left(\frac{33000}{3300} \right)$$

$$= \left(\frac{1}{0.0038} \right) \times 10$$

$$= 263.16 \times 10$$
$$= 2,631.6\ \text{Hz}$$

Decreasing the value of either resistor R3 or capacitor C1 (or both) increases the output signal frequency, and vice versa.

Another simple two-waveform op amp function-generator circuit is shown in Fig. 4-8. This one generates a rectangle wave and a somewhat modified sawtooth wave. Both waveforms are available simultaneously. Both always have the same frequency.

This circuit should look rather familiar to you. It is really nothing more than one of the simple op amp rectangle-wave generator circuits introduced back in chapter 3. There is no difference in the actual circuitry. The only difference here is that a second output is tapped off across the timing capacitor (C1).

The capacitor is charged on the positive half-cycles and discharged on the negative half-cycles, so the voltage across it is a sort of modified sawtooth wave, as illustrated in Fig. 4-9. As you can see here, this isn't a very good sawtooth wave at all. In fact, it is more like a cross between a sawtooth wave and a triangle wave. This is because the discharge time is relatively long. For a true sawtooth wave, most of each

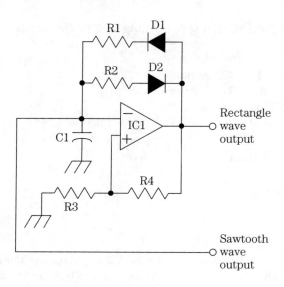

4-8 This alternate simple two-waveform op amp function-generator circuit produces a rectangle wave and a somewhat modified sawtooth wave.

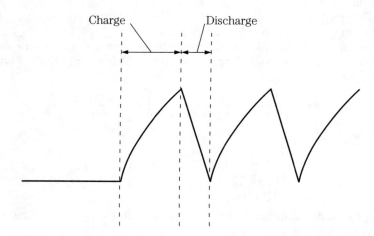

4-9 The capacitor is charged on the positive half-cycles and discharged on the negative half-cycles, so the voltage across it is a sort of modified sawtooth wave.

cycle should be devoted to charging the capacitor. The closer the discharge period is to instantaneous, the better. This simple circuit can not generate a true sawtooth wave, but it can come quite close, and should be good enough for many non-critical practical applications.

The duty cycle of the rectangle wave determines how good the sawtooth wave will be. For this application, a narrow LOW pulse is desired. The duty cycle is set by the ratio of the values of resistors R1 and R2. A ratio of at least 1:10, and preferably larger, should be used in this application.

To get a better sawtooth-wave generator from this basic circuit, you can modify it, as shown in Fig. 4-10. This circuit produces a much more linear output. It is similar to the previous circuit, except this time, the feedback resistors and diodes are replaced by a FET (field-effect transistor) Q1. Resistor R1 biases the FET to pinch-off.

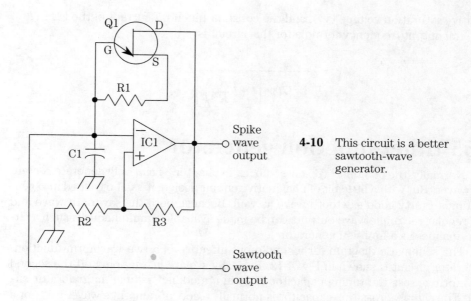

4-10 This circuit is a better sawtooth-wave generator.

Capacitor C1 is charged linearly during the positive half-cycle, because the FET functions as a constant-current source. When the voltage across the capacitor exceeds the positive feedback voltage (at the op amp's non-inverting input), the op amp switches to negative saturation. The FET, which is in the inverting input feedback path, now electrically looks like a forward-biased diode—allowing capacitor C1 to discharge very rapidly. Once this capacitor is discharged sufficiently, the op amp switches back to its positive saturation condition, and a new cycle begins.

While this circuit generates a much better sawtooth wave, the second waveform, taken directly from the op amp's output, will no longer be a rectangle wave at all. Instead, the signal appearing at this output will be a rather odd negative spike waveform, as illustrated in Fig. 4-11. Each time the op amp switches to the negative saturation state, a brief negative spike appears at the output. The op amp does not remain in negative saturation for long—just a tiny fraction of each complete cycle. The rest of the time, the op amp's output remains at the positive saturation voltage.

Calculating the output signal frequency for this circuit is a bit awkward. Of course, both output waveforms will always have exactly the same frequency. To calculate the frequency, you must use the manufacturer's spec sheets for the semiconductor components you are using in your circuit. You must know the op amp's

4-11 The second output from the circuit of Fig. 4-10, which is taken directly from the op amp's output, is a rather odd negative spike waveform.

positive saturation voltage (V_s), and the constant pinch-off current of the FET (I_{ds}). The full output frequency formula for this circuit is:

$$F = \frac{I_{ds}}{2V_s C_1 \times \left[I + \left(\dfrac{R_3}{R_2}\right)\right]}$$

555 timer function generator

Normally, the 555 timer (discussed back in chapter 3) can only generate rectangle waves. But with a little bit of imaginative circuit design, it can be tricked into generating a pretty good sawtooth wave as well. Because both the sawtooth wave and the regular rectangle wave outputs can be made available simultaneously, such a circuit qualifies as a function generator.

The schematic diagram for a simple-but-effective 555 two-waveform function-generator circuit is shown in Fig. 4-12. Basically, we are just tapping off the second waveform across the timing capacitor (C1) in a modified astable multivibrator circuit. The signal across this capacitor is normally a sort of ramp-like wave. It is not a perfectly linear ramp, however. It does have some curve to it.

If the duty cycle of the rectangle wave is set up close to 1:2, the signal across the timing capacitor becomes a distorted triangle wave, as illustrated in Fig. 4-13. At the other extreme, if the duty cycle is made 1:10 or better, the signal across the timing capacitor resembles a marginally-distorted ascending sawtooth wave, as illustrated in Fig. 4-14.

Taken directly from across the capacitor, these waveforms will not be very linear, and there can also be significant problems with loading effects. In most cases, a buffer amplifier of some type will probably be necessary in most practical applications. That is why the circuit of Fig. 4-12 includes a couple of transistors—to limit potentially problematic loading effects and to increase the linearity of the added output signal.

Transistor Q1 serves as a constant current source to charge the capacitor linearly. The timing period for the charging half-cycle is:

$$T_1 = \frac{\left(\dfrac{C_1 V_{CC}}{3}\right)}{I_t}$$

where V_{cc} is the circuit's supply voltage (in volts), C_1 is the timing capacitance (in farads), and I_t is the output current from Q1 in amperes. This current will be set by

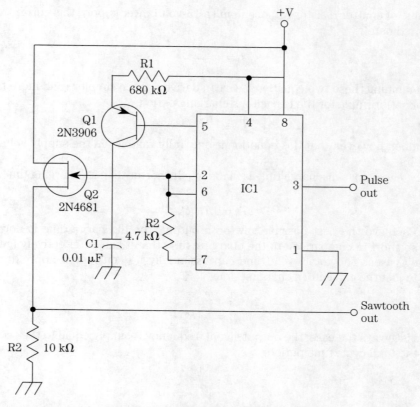

4-12 This circuit is a simple but effective 555 two-waveform function generator.

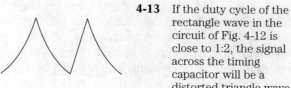

4-13 If the duty cycle of the rectangle wave in the circuit of Fig. 4-12 is close to 1:2, the signal across the timing capacitor will be a distorted triangle wave.

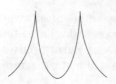

4-14 If the duty cycle of the rectangle wave in the circuit of Fig. 4-12 is 1:10 or better, the signal across the timing capacitor will be a marginally distorted ascending sawtooth wave.

the value of emitter resistor R3. Assuming a 15-volt power supply, this current value is approximately:

$$I_t = \frac{4.4V_{cc}}{R_3}$$

Combining these two equations, we can derive a much simpler (and more familiar-looking) formula for the capacitor's charging period:

$$T_1 = 1.1R_3C_1$$

Remember, however, that this equation is only fully valid when the supply voltage is +15 volts.

The capacitor's discharge time is made much shorter than its charging time. The formula is:

$$T_2 = 0.7R_1C_1$$

To achieve the best possible sawtooth output, the discharge time is normally made so short in comparison to the charging time that it can be effectively ignored in many cases. The total cycle time can essentially be considered approximately equal to just the capacitor's charging time:

$$T_t = T_1$$
$$= 1.1R_3C_1$$

As is always the case, the output signal frequency is simply equal to the reciprocal of the total cycle time period:

$$F = \frac{1}{T_t}$$

With a little bit of algebraic rearranging, we can solve for the output frequency directly from the component values:

$$F = \frac{0.91}{(R_3C_1)}$$

Remember, this equation is only an approximation. It ignores the effects of the brief discharge period (T_2). In most applications, this formula will be close enough, but it might not be adequate in certain precision applications.

This circuit will function with a supply voltage of less than +15 volts, but the equations presented here become increasingly inaccurate as the supply voltage is decreased.

Transistor Q2 is a *FET* (*field-effect transistor*). Its job in this circuit is to act as a buffer amplifier for the sawtooth wave output, preventing loading problems that would be inevitable if the signal was tapped directly off across the timing capacitor (C1).

The normal rectangle wave output signal is simultaneously available at pin #3 of the 555 timer chip, as usual. Because of the very short discharge time necessary for the sawtooth wave, this output will be a string of very narrow pulses.

Using the component values suggested in the schematic diagram, the output signal frequency will be about 100 Hz.

UJT VCO

A *UJT*, or *unijunction transistor*, is a very good sawtooth-wave generator. As we shall see, a couple additional waveforms can also be tapped off, making it a function generator.

Usually when we refer to a *transistor*, we mean a bipolar transistor, which is constructed like sort of a semiconductor sandwich, as illustrated in Fig. 4-15. A pure semiconductor crystal can be doped with certain impurities to create an excess of electrons (*n-type semiconductor*), while different impurities will create a shortage of electrons, or excess of "holes" (*p-type semiconductor*). In an npn bipolar transistor, there are two relatively large sections of n-type semiconductor separated by a rather thin section of p-type semiconductor. In a pnp bipolar transistor, the polarities are simply reversed—so we have two relatively large sections of p-type semiconductor separated by a rather thin section of n-type semiconductor.

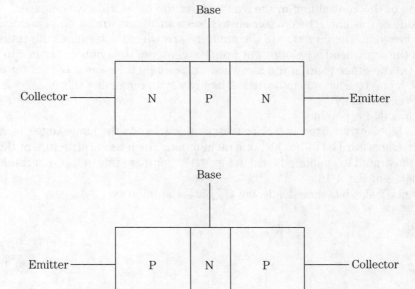

4-15   A standard bipolar transistor is constructed like a semiconductor sandwich.

A standard bipolar transistor (whether npn or pnp) has three leads identified as follows:

- Collector
- Base
- Emitter

The standard schematic symbols for npn and pnp bipolar transistors are shown in Fig. 4-16. The three leads are sometimes labelled on schematics, but they usually aren't. They are easy enough to recognize. The collector and emitter are always on

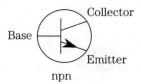

npn

4-16 The standard schematic symbols for npn and pnp bipolar transistors.

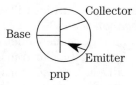

pnp

one side of the central line in the symbol, and the base is always centered on the other side of this line. The emitter always has a small arrowhead on it, and the collector does not. The direction of the emitter's arrowhead determines the transistor type. If the arrowhead is pointing out from the center of the symbol, it is an npn transistor. On the other hand, if the arrowhead is pointing in towards the center of the symbol, a pnp transistor is indicated. A handy way to remember this is:

Never **P**oints i**N**

Points i**N** **P**erpetually

In a bipolar transistor, there are two pn junctions. As the name suggests, a unijunction transistor (UJT) has just one pn junction. The internal structure of this device is illustrated in simplified form in Fig. 4-17. Compare this to the semiconductor sandwiches of Fig. 4-16.

The UJT also has three leads, but they are identified as:

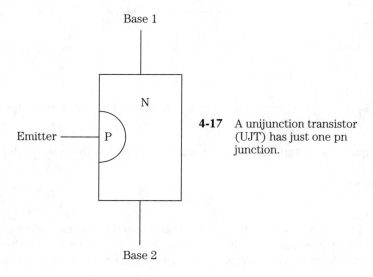

4-17 A unijunction transistor (UJT) has just one pn junction.

Emitter   E
Base 1   B1
Base 2   B2

There is no collector on a UJT, but there are two separate base connections. The schematic symbol for a UJT is shown in Fig. 4-18. The emitter has a small arrowhead on it, as in the symbol for a standard bipolar transistor. The bases do not. In most applications, the bases are functionally interchangeable, so we generally don't need to distinguish between them. Where it matters, they will usually be explicitly labelled in the schematic diagram. While it's not a hard and fast rule, it is the common practice to consider the uppermost base in the diagram as "base 1", and the lower one as "base 2."

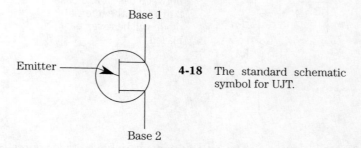

4-18   The standard schematic symbol for UJT.

Most UJTs on the market today are of the n-type, with the arrowhead on the schematic symbol pointing inward. p-type UJTs do exist, but they are uncommon, and tend to be rather expensive, as well as difficult to find.

Figure 4-19 shows a very rough equivalent circuit for a UJT. Electrically, the large central n-type semiconductor section acts like a resistive voltage divider, with a diode (the component's single pn junction) connected to the common ends of the two resistive elements. In an equivalent circuit for a p-type UJT, the polarity of the diode would simply be reversed.

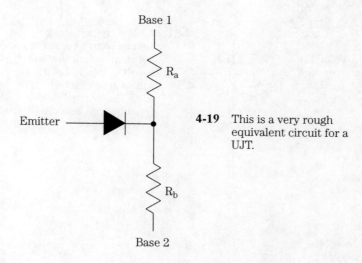

4-19   This is a very rough equivalent circuit for a UJT.

Unless stated otherwise, the emitter of a UJT is assumed to be at ground potential (0 volts). If a voltage is applied between base 1 and base 2, as shown in Fig. 4-20,
the internal diode will be reverse-biased. Of course, this means that no current can
flow from the emitter to either of the base terminals.

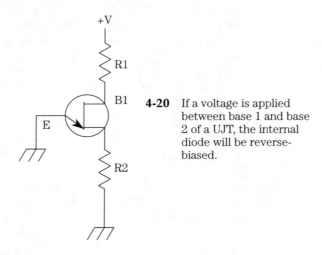

4-20 If a voltage is applied
between base 1 and base
2 of a UJT, the internal
diode will be reverse-
biased.

Now, let's suppose there is also an additional variable voltage source connected
between the emitter and base 1, as illustrated in Fig. 4-21. As this emitter-base 1
voltage is increased from zero, a point will soon be reached when the internal diode
(pn junction) becomes forward-biased. Beyond this point, current can flow from the
emitter to both the bases. The UJT is turned-on when it's internal pn junction is forward-biased, and turned-off when the pn junction is reverse-biased.

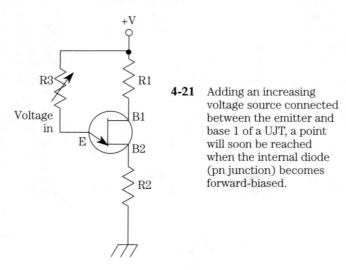

4-21 Adding an increasing
voltage source connected
between the emitter and
base 1 of a UJT, a point
will soon be reached
when the internal diode
(pn junction) becomes
forward-biased.

For signal generation, external pulses are usually fed into the emitter of a UJT. Assuming the input pulses have sufficient amplitude to forward-bias the pn junction, output pulses will appear at the two bases, and can be tapped off across either resistor R1 or R2, or both. The signal tapped off across resistor R2 will have the opposite polarity as the signal tapped off across resistor R1. Naturally, the output pulses will be in step (in phase) with the input pulses.

UJTs are most commonly used in timer circuit and oscillator circuits. These two applications are functionally very similar to one another, but for our purposes in this book, we are only interested in the oscillator circuits.

Because input pulses must be externally fed into the emitter of the UJT to get output pulses, it would seem like you'd need a complete oscillator circuit without the UJT just to get the UJT to work. So what is this component good for? Actually, the UJT can generate its own input pulses through an external RC (resistor–capacitor) network.

UJTs are very versatile devices in practical oscillator circuits. They can be used reliably at frequencies as low as 1 Hz, and as high as 1 MHz (1,000,000 Hz) (occasionally even higher with some specialized devices), depending on the values of the external passive components used in the oscillator circuit. This is quite an impressive operating range, especially considering how simple the basic UJT oscillator circuit is.

The simplest way to use a UJT for oscillator applications is in a *relaxation oscillator circuit*. The basic form of this circuit is illustrated in Fig. 4-22. Basically, we have just added the frequency-determining RC network to the conceptual UJT demonstration circuit of Fig. 4-21.

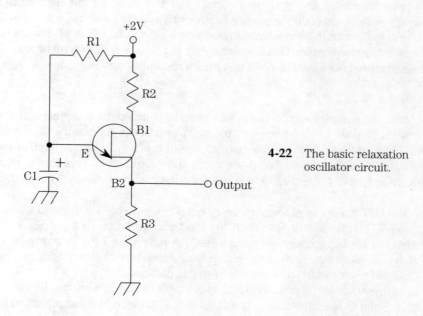

4-22 The basic relaxation oscillator circuit.

The power supply voltage to a UJT circuit is often called V_{bb} in technical literature. This is because it is applied across the device's two bases. We will just refer to the supply voltage as V+ to keep things simple. Usually, only a very low supply voltage is required to get a UJT circuit to function.

When the supply voltage is first applied to this simple relaxation oscillator circuit, the voltage across capacitor C1 is obviously 0 volts, because the capacitor hasn't had a chance to be charged at all yet. The instant power is applied to the circuit, capacitor C1 will start to charge up through resistor R1. The capacitor will try to charge up to a voltage equal to the supply voltage.

The voltage across this capacitor is directly applied to the emitter of the UJT, so the emitter voltage will always be equal to the present charge on this timing capacitor. At some point, the voltage across the capacitor will exceed the internal turn-on point of the UJT, forward-biasing its internal pn junction. An output pulse will be generated to the two base terminals. This is often referred to as the *UJT firing*.

When the UJT fires, it quickly discharges the capacitor. The capacitor's stored voltage is drained off through the UJT's emitter. In a very short time, the voltage across the capacitor is not sufficient to forward-bias the UJT's internal pn junction, so the device is turned off. Now the capacitor starts charging up through resistor R1 again, and the entire process is repeated. This cycle is repeated as long as power is applied to the circuit, making this an oscillator circuit.

The time it takes capacitor C1 to charge up enough to fire the UJT is dependent on several component values throughout the circuit. The most important of these are the value of capacitor C1 itself, resistor R1, and a specification known as the intrinsic stand-off ratio of the UJT. This last specification can be found in the manufacturer's spec sheet for the particular device used in your circuit. Fortunately, except in very critical, precision applications, the intrinsic stand-off ratio can usually be ignored altogether. In most general-purpose application, the effect of this specification is essentially negligible. This leaves us with capacitance C_1 and resistance R_1. The charging time of the capacitor can be considered approximately equal to:

$$T_c = C_1 R_1$$

This formula is hardly exact, but when normal component tolerances are assumed, it will probably be close enough for most practical applications. The resistance can be made variable with a potentiometer or a trimpot, so the output frequency can be precisely adjusted if the application requires it. This timing equation is still a handy tool for getting you in the right approximate range of component values.

When the UJT fires, the capacitor is very quickly discharged to ground through base 2 of the UJT and resistor R3. Then the next charging cycle then begins. The value of resistor R3 is typically very small, so this discharge time is normally extremely short. In most practical applications, it is considered quite negligible, and can be reasonably ignored in the circuit design calculations.

For most practical purposes, the total cycle time in this circuit can be considered roughly equal to the capacitor's charging time. Therefore, the output signal frequency is approximately equal to:

$$F = \frac{1}{T_c}$$

$$= \frac{1}{(C_1 R_1)}$$

This is just about as simple as electronics math ever gets.

In most common oscillator circuits, it is most practical to select a likely capacitor value and then solve for the necessary resistance to obtain the desired output frequency. This is usually the best approach because resistors are almost always readily available in more varied sizes than capacitors, and can easily be made adjustable by using a potentiometer or a trimpot.

In the case of a UJT oscillator circuit, however, this usual recommended procedure is best reversed. Select a suitable resistor value, and then solve for the necessary capacitance. You can fine-tune adjust the resistance later. This reverse design process is used because this type of circuit places special requirements on the resistor value, Ideally, the timing resistor (R1 in our schematic) should have a value between:

$$R_{1\min} = \frac{[2(V_{bb} - V_v)]}{I_v}$$

and:

$$R_{1\max} = \frac{[V_{bb}(1-\eta) - 0.5]}{2I_p}$$

Don't let these ominous-looking equations and all their strange symbols throw you. For the most part, these values are all taken directly from the manufacturer's spec sheet for the specific UJT you are using.

V_{bb}, of course, is simply the circuit's supply voltage, or $V+$. I_p is the maximum current that can be permitted to flow from the emitter to base 1. V_v is the *valley voltage*, or the voltage from the emitter to base 1 in the instant just after the UJT has started to conduct. I_v is the *valley current* which exists under the same conditions as V_v. η is the intrinsic stand-off ratio, and its value is determined by the internal base resistances of the UJT (refer back to Fig. 4-19):

$$\eta = \frac{R_a}{(R_a + R_b)}$$

In most practical UJTs, the value of Ra will be close to that of Rb, so the intrinsic stand-off ratio will typically be quite small—usually about 0.5 to 0.8.

This might seem to be terribly confusing, but it shouldn't be so bad, once we work our way through a typical design example. Let's design a UJT relaxation oscillator circuit with an output frequency of 1.7 kHz (1,700 Hz). The output signal amplitude across resistor R3 should be 2.5 volts peak-to-peak. We will use a +12-volt supply voltage to power our circuit.

First, we check the manufacturer's spec sheet to get the relevant internal values. We will assume our UJT has the following specifications:

$$\eta = 0.57$$
$$I_p = 10 \ \mu A \qquad (0.00001 \ \text{ampere})$$

$$V_v = 3.2 \text{ volts}$$
$$I_v = 25 \text{ mA} \qquad (0.025 \text{ ampere})$$
$$\text{Total internal resistance } (R_a + R_b) = 9 \text{ k}\Omega \qquad (9{,}000 \text{ ohms})$$

These values are all fairly typical of most common UJTs.

Using these specifications, we can determine the minimum acceptable value for timing resistor R1:

$$R_{1min} = \frac{[2(V_{bb} - V_v)]}{I_v}$$

$$= \frac{[2(12 - 3.2)]}{0.025}$$

$$= \frac{(2 \times 8.8)}{0.025}$$

$$= \frac{17.6}{0.025}$$

$$= 704 \text{ ohms}$$

On the other hand, the maximum acceptable value for this component is:

$$R_{1max} = \frac{[V_{bb}(1 - \eta) - 0.5]}{2I_p}$$

$$= \frac{[12(1 - 0.57) - 0.5]}{(2 \times 0.00001)}$$

$$= \frac{(12 \times 0.43 - 0.5)}{0.00002}$$

$$= \frac{(5.16 - 0.5)}{0.00002}$$

$$= \frac{4.66}{0.00002}$$

$$= 233{,}000 \text{ ohms}$$
$$= 233 \text{ k}\Omega \ (233{,}000 \text{ ohms})$$

As you can see, the value of R1 really isn't limited all that much. There is quite a wide range of acceptable resistance values, but you must never permit this resistance to go beyond these limits, or the circuit will not work properly, and the UJT could be damaged or destroyed.

You could make the circuit manually adjustable over a wide range of frequencies using this hypothetical UJT, by replacing resistor R1 with a 200-kΩ potentiometer in series with a 820-Ωohm or 1-kΩ resistor. The fixed series resistor sets the minimum resistance value, and the maximum is simply the sum of the potentiometer's full value and the fixed resistance (201 kΩ). No matter how we adjust the potentiometer in this example, the total effective resistance will be within the acceptable limits. However, a 250-kΩ potentiometer could be set for a potentially damaging, unacceptable resistance value. Be careful.

Let's select a convenient value somewhere in the midrange of the acceptable resistance values so we can solve for the timing capacitor value (C1). 100 kΩ (100,000 ohms) would be a nice mid-range value that will be easy to work with.

Rearranging the approximate frequency formula given earlier in this section, we can solve for the unknown capacitance:

$$C_1 = \frac{1}{FR_1}$$

$$= \frac{1}{(1{,}700 \times 100{,}000)}$$

$$= \frac{1}{170{,}000{,}000}$$

$$= 0.0000000059 \text{ farad}$$

$$= 0.0059 \text{ }\mu\text{F}$$

Now, 0.0059 μF is not a standard capacitor value, but 0.0047 μF should be close enough. If we want to get a little more precise, we can recalculate the value of R1 using the rounded-off capacitor value. We started out with a resistance value near the acceptable range, and have changed the capacitance value only slightly from its nominal calculated value, so we have no worry of coming up with an unacceptable resistance value:

$$R_1 = \frac{1}{FC_1}$$

$$= \frac{1}{(1700 \times 0.0000000059)}$$

$$= \frac{1}{0.00001003}$$

$$= 99{,}701 \text{ ohms}$$

You can see how little difference rounding off the capacitor made. When we round off this calculated resistance value to the nearest standard resistor value, we again get 100 kΩ (100,000 ohms). In some practical applications, things might not work out quite this neatly, but they should be reasonably close.

Now, we need to figure out what values to use for base resistors R2 and R3. The equation for finding the desired value of resistor R2 is:

$$R_2 = \frac{(R_a + R_b)}{(2 \, \eta \, V_{bb})}$$

We know the supply voltage (V_{bb}) is 12 volts. The other variables are taken from the manufacturer's spec sheet for the particular UJT being used in the circuit. For our hypothetical UJT in this example, the equation works out to:

$$R_2 = \frac{9000}{(2 \times 0.57 \times 12)}$$

$$= \frac{9000}{13.68}$$

$$= 658 \text{ ohms}$$

A standard 680-ohm resistor should be close enough. Notice that this resistance will almost always be fairly small, typically less than 1 kW (1,000 ohms).

This leaves just resistor R3. Our output signal is tapped off across this resistor, and earlier we stated we wanted an output signal amplitude of 2.5 volts peak-to-peak. For the purpose of calculating the value of this resistor, we only need to bother with the positive-peak voltage, which is +2.5 volts in this case.

The circuit diagram is redrawn conceptually in Fig. 4-23. Notice that R3 is part of a simple resistive voltage divider network. We know the supply voltage (12 volts) and the desired voltage across this resistor (2.5 volts), and all the other resistor values in the string (R2 = 680 ohms, the series combination of R_a and R_b is 9000 ohms).

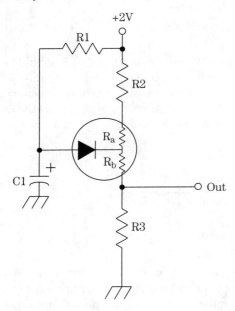

4-23 This redrawn version of the basic relaxation oscillator circuit shows that R3 is part of a simple resistive voltage-divider network.

The total resistance of the string is equal to:

$$R_t = R_2 + (R_a + R_b) + R_3$$
$$= 680 + 9000 + R_3$$
$$= 9680 + R_3$$

The voltage drop across R2 and $(R_a + R_b)$ must add up to the supply voltage less the desired voltage drop across R3:

$$V_d = V_{bb} - V_{r3}$$
$$= 12 - 2.5$$
$$= 9.5 \text{ volts}$$

We can use Ohm's law to find the current flow through the upper portion of the voltage divider string (everything except resistor R3):

$$I = \frac{E}{R}$$

$$= \frac{9.5}{9680}$$

$$= 0.0009814 \text{ ampere}$$

Because R3 is in series with these other resistances, there must be the same current flowing through this resistor too. So we can rearrange the Ohm's law equation, and solve for the desired resistance value:

$$R_3 = \frac{E}{I}$$

$$= \frac{2.5}{0.0009814}$$

$$= 2,547 \text{ ohms}$$

A 2.7-kΩ (2,700 ohms) resistor should be close enough for most practical purposes. If the output signal is really critical, we can make up R_3 of a 1-kΩ (1,000 ohms) trimpot in series with a 1.8-kΩ (1,800 ohms) resistor. We can then calibrate the trimpot to give the exact desired output amplitude. This would be easiest to do by monitoring the output signal's amplitude on a good oscilloscope.

So far, we haven't mentioned the output waveform generated by this circuit. It is a rather odd one—a string of positive spikes, as illustrated in Fig. 4-24.

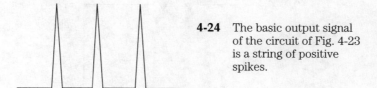

4-24 The basic output signal of the circuit of Fig. 4-23 is a string of positive spikes.

We can tap a second output signal off across resistor R2. This will be identical to the signal across resistor R3, except that all polarities will be reversed, giving us a series of negative spikes across this resistor, as shown in Fig. 4-25. These two signals will be in phase with each other. When there is a positive spike across resistor R3, at that same instant there will be a negative spike across resistor R2. The signal amplitude will also be different, because the resistance of R2 is not the same as that of R3, but the same current will always flow through both these resistors. Therefore, the voltage dropped across each resistor must be different.

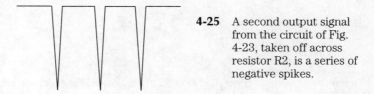

4-25 A second output signal from the circuit of Fig. 4-23, taken off across resistor R2, is a series of negative spikes.

A more useful waveform can be tapped off across the timing capacitor (C1). This is an ascending sawtooth wave, as shown in Fig. 4-26. Unfortunately, as the circuit is shown here, this is the most unreliable waveform for practical use. An external output circuit or device is likely to have loading effects on the capacitor, affecting the overall operation of the circuit. Such loading problems can be avoided by adding a buffer amplifier stage between the capacitor and the output circuit or device. This will be done in our next, more sophisticated version of this circuit.

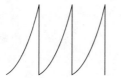

4-26 A more useful waveform (an ascending sawtooth wave) can be tapped off across the timing capacitor of the circuit in Fig. 4-23.

An improved version of this basic UJT relaxation oscillator circuit is shown in Fig. 4-27. This circuit is actually a *VCO*, or *voltage-controlled oscillator*. The component values determine the base frequency in the usual way, as described above. But an external voltage input can detune the oscillator frequency in a predictable

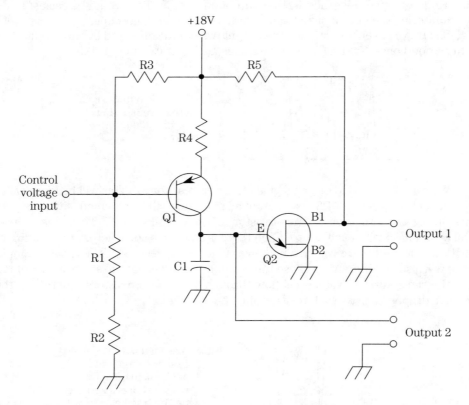

4-27 An improved, voltage-controlled version of the basic UJT relaxation oscillator circuit.

way. The output signal frequency will change proportionately in response to any changes in the input voltage.

In addition to being a VCO, this circuit is still a function generator, because it has three different outputs, each with its own output. These are the same waveshapes we had in the earlier, simpler version of this UJT circuit:

Output 2   Ascending sawtooth wave
Output 1   Positive spike wave

All three output waveforms will always have the same frequency.

The addition of transistor Q2 to the circuit permits the use of external voltage control. It also buffers the sawtooth output waveform taken off across the timing capacitor (C1), limiting adverse loading effects. This output signal is taken off from Q2's collector. Q2 is a standard pnp bipolar transistor, not a UJT.

A typical parts list for this circuit is given in Table 4-1. None of the component values are particularly critical. Feel free to substitute nearby values if you choose.

**Table 4-1. Typical parts list for the
UJT VCO circuit of Fig. 4-27**

Q1	Pnp transistor (2N1309, or similar)
Q2	UJT (2N491, or similar)
C1	0.1-µF capacitor (see text)
R1	330-kΩ, ¼-W, 5% resistor
R2	10-kΩ, ¼-W, 5% resistor (see text)
R3, R5	330-Ω, ¼-W, 5% resistor)
R4	1-MΩ, ¼-W, 5% resistor)

The components of most interest to us here are resistor R2 and capacitor C1, which determine the base output frequency. This is the signal frequency that will be generated when the control voltage input is 0. The larger these two component values are, the lower the base output frequency will be. For best results, resistor R2 should have a value of at least 1 kΩ (1,000 ohms), and no more than 22 kΩ (22,000 ohms). Similarly, the recommended range of values for capacitor C1 is 0.001 µF (0.000000001 farad) to 10 µF (0.00001 farad).

The formula for the output frequency from this circuit is:

$$F = \left(\frac{2}{R_2 C_1}\right) \times \left[1 - \left(\frac{V_i}{V_{bb}}\right)\right]$$

where R_2 is the resistance in ohms, C is the capacitance in farads, V_{bb} is the circuit's supply voltage in volts, and V_i is the control voltage input, also in volts. The resulting frequency will be in Hertz.

This particular circuit works best with supply voltages between 18 volts and 24 volts, and input control voltages in the 10- to 20-volt range. The input voltage should always be less than the supply voltage, or the transistors in the circuit could be damaged or destroyed. In our examples, we will use the component values suggested in the parts list, and we will assume a supply voltage of 22.5 volts.

First, let's assume that the control voltage at the input is 10 volts. In this case, the output signals should have a frequency of about:

$$F = \left[\frac{2}{(10000 \times 0.0000001)} \right] \times \left[1 - \left(\frac{10}{22.5} \right) \right]$$

$$= \left(\frac{2}{0.001} \right) \times (1 - 0.444)$$

$$= 2000 \times 0.556$$

$$= 1111 \text{ Hz}$$

Now, let's increase the control voltage to 15 volts. This time we get an output signal frequency of approximately:

$$F = \left[\frac{2}{(10000 \times 0.0000001)} \right] \times \left[1 - \left(\frac{15}{22.5} \right) \right]$$

$$= \left(\frac{2}{0.001} \right) \times (1 - 0.667)$$

$$= 2000 \times 0.333$$

$$= 666 \text{ Hz}$$

One last example. This time we will raise the input control voltage all the way to 20 volts, giving us an output signal frequency of about:

$$F = \left[\frac{2}{(10000 \times 0.0000001)} \right] \times \left[1 - \left(\frac{20}{22.5} \right) \right]$$

$$= \left(\frac{2}{0.001} \right) \times (1 - 0.889)$$

$$= 2000 \times 0.111$$

$$= 222 \text{ Hz}$$

Increasing the control voltage, decreases the output frequency, and vice-versa.

You can make this circuit more versatile by adding a manual-base frequency control. Simply replace resistor R2 with a 20-kΩ potentiometer in series with a 1-kΩ fixed resistor. You can also add over-lapping frequency ranges by using a rotary switch to select one of several paralleled timing capacitors in place of C1.

Increasing the value of either resistor R2 or capacitor C1 will decrease the base output frequency of this circuit.

Dedicated function-generator ICs

A number of dedicated function-generator ICs have appeared on the market over the last few years. These devices can generate very pure signals of the basic waveforms, and sometimes produce some unusual waveshapes as well. Usually, a function-generator IC can put out signal frequencies over a very wide range, typically extending well beyond the audible frequency spectrum, especially on the high end.

Two of the most popular function-generator ICs now on the market are the 8038 and the XR 2206. Of course, there is no way to guarantee that either or both of these devices won't become discontinued and obsolete without notice between the time this is written and when you read it. Before investing any money into buying components for any electronics project, always make sure you can find a source for any critical ICs, or other potentially difficult-to-find components. It can be very frustrating to gather together everything else you need for a given circuit only to find an essential IC or other specialized component is no longer available. Unfortunately, neither the author nor the publisher have the resources to keep track of specific commercial sources for electronic components.

At any rate, the two specific chips we will be discussing in this section can be considered fairly typical of what can be expected in modern function-generator ICs.

The 8038 has long been a favorite among circuit designers and electronics hobbyists. It is quite powerful, versatile, and fairly easy to work with in practical circuits. A pin-out diagram of this chip is shown in Fig. 4-28.

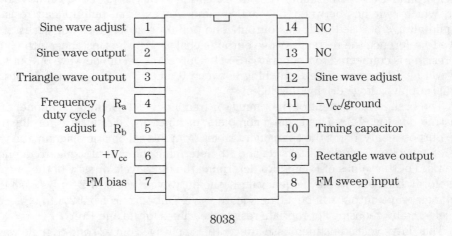

8038

4-28 The 8038 is a popular function-generator IC.

The 8038 is designed to generate all four of the basic waveforms—sine waves, rectangle waves, triangle waves and sawtooth waves. The rectangle waves generated by this device can be adjusted to almost any desired duty cycle. This chip has an incredible frequency range. It can generate signals with frequencies ranging from 0.001 Hz (one cycle every 1000 seconds, or every 16.667 minutes) to above 1 MHz (1,000,000 Hz). Typically, the output waveshapes will have no more than 1.0% distortion. The linearity is specified as better than 0.1%. The output signal frequency of the 8038 exhibits very little drift. The spec sheet claims the frequency drift is no more than 50 ppm (parts per million) per degree centigrade. Any frequency drift effects are more likely to be the fault of the external components used in the circuit rather than in the 8038 chip itself.

The 8038 function generator can also be used as a *VCO* (*voltage-controlled os- cillator*), with an external control voltage signal at the FM input (pin #8), adjusting the output signal frequency within a specified range.

The nominal output frequency (ignoring any input control voltage) can be set with two external timing resistors and one timing capacitor. A wide frequency range can be achieved with a single capacitor value.

Even the power supply requirements for the 8038 are impressively flexible. Ei- ther a dual-polarity power supply or a single-polarity power supply can be used with this chip. For a dual-polarity power supply, anything from ±5 volts to ±15 volts is ac- ceptable. The usable range for single-ended supply voltages ranges from +10 volts up to +30 volts.

When using a dual-polarity power supply, the output signal will be centered around true ground potential (0 volts). If a single-ended supply voltage is used with the 8038, the output signal will be symmetrical around one-half the supply voltage. For example, if a +18-volt power supply is used, the output waveforms will be cen- tered around a +9-volt baseline. The one exception to this is the rectangle wave out- put, which switches between $+V_{cc}$ and ground. An external pull-up/load resistor should always be used with this output. The pull-up resistor does not necessarily have to reference the rectangle wave output signal to $+V_{cc}$. For instance, if the pull- up resistor is connected to a separate +5-volt source, the rectangle wave output sig- nal will be TTL-compatible (switching between 0 and +5 volts), regardless of the actual supply voltage driving the 8038.

The basic arrangement of the timing, or frequency-determining components is illustrated in Fig. 4-29. This is not a complete, or functional circuit. It is for illustra- tive purposes only. Using the component labels from this schematic diagram, the val- ues of capacitor C, and resistors Ra and Rb determine the output signal frequency. The ratio of the values of Ra and Rb determine the duty cycle or slew of the output waveform. If Ra and Rb have equal values, the duty cycle will be 1:2, or 50%, so the rectangle wave output will be a true square wave. Duty cycles from 2% to 99% can be selected by choosing appropriate resistance values for Ra and Rb.

This duty cycle adjustment also affects the sine wave (pin #2) and triangle wave (pin #3) outputs. For a 50% duty cycle ($R_a = R_b$), these pins will put out the standard waveforms they are named for. As the duty cycle is moved further from 50% (in ei- ther direction), these waveshapes will become increasingly distorted. At the ex- tremes of the duty cycle range, the triangle wave will generate a sawtooth wave. This is why there is no dedicated output pin for this waveform. Pin #3 does double duty.

Potentiometers can be used for R_a and R_b, but this won't be very practical in most real applications. Adjusting the circuit for a specific desired frequency and duty cycle would be difficult because the two controls interact so much, each affect- ing both the output signal frequency and the duty cycle. A solution to this problem is illustrated in Fig. 4-30. In this circuit, the operator has full control over both the frequency and the duty cycle, independently via the two potentiometers. R1 controls the frequency but does not affect the duty cycle, while R2 can change the duty cycle without affecting the frequency.

Potentiometer R2 can be eliminated from the circuit, for a fixed-duty cycle set by the relative values of resistors Ra and Rb as before. Making these two resistors

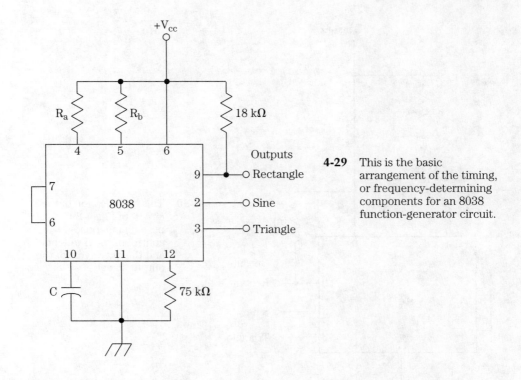

4-29 This is the basic arrangement of the timing, or frequency-determining components for an 8038 function-generator circuit.

equal in value for a 50% duty cycle is usually the best choice, except in applications requiring a sawtooth wave output.

If you will only need a fixed 50% duty cycle (the available outputs will be sine wave, triangle wave and square wave), you can simplify the circuitry as shown in Fig. 4-31. There is no way to adjust this version of the circuit for any duty cycle other than 50%. This is the simplest approach to frequency control for the 8038 function generator.

A practical audio-frequency function generator circuit using the 8038 is illustrated in Fig. 4-32. The output frequency from this circuit can be varied over a 1000:1 range, covering the entire audio frequency spectrum (20 Hz to 20 kHz). This wide range is achieved by applying different dc control voltages to the FM sweep input (pin #8) via potentiometer R4. Meanwhile, the voltage across the timing resistors (R1 and R3) is held at a relatively low level by the 1N457 diode. The voltage supplied to the timing resistors and the duty-control potentiometer (R3) is several millivolts below the maximum voltage ($+V_{cc}$) available to the frequency-control potentiometer (R4).

R6 is a trimpot, not a front panel control. It is used for calibration of the circuit only—a "set and forget" control. It is adjusted to minimize variations in the output signal's duty cycle with changes in frequency. Trimpot R7 is adjusted for minimum sine wave distortion at a 50% duty cycle. Both of these calibration adjustments are best made with the aid of a good oscilloscope.

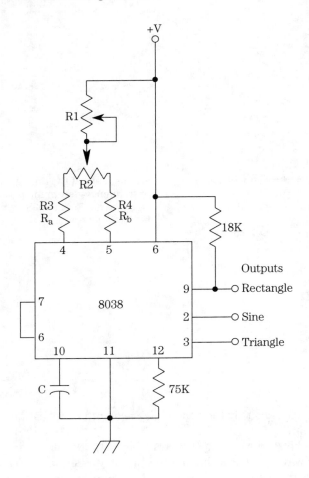

4-30 This modification of the circuit of Fig. 4-29 permits independent control of the frequency and the duty cycle of the output signal.

Once calibrated, this audio-frequency function generator will exhibit excellent specifications, comparable to many professional function generators used for serious electronics testing.

Another popular and versatile function-generator IC is the XR 2206. We have already used this device in the audio frequency sine-wave oscillator project of chapter 2. We will now look at this chip in a more general and thorough way.

Like the 8038, the XR 2206 generates high-quality (low distortion) sine waves, triangle waves, sawtooth waves and rectangle waves of various duty cycles, although it uses different means to accomplish these same ends. The output frequency range is very wide, running from a few fractions of hertz up to several hundred kilohertz (1 kilohertz = 1,000 hertz). For a given set of component values, a single variable resistance or external control voltage can vary the output frequency of the XR 2206 over a 2000:1 range. The pin-out diagram for the XR 2206 function generator is shown in Fig. 4-33.

The power supply requirements for this device are quite flexible. Either a dual-polarity power supply or a single-polarity power supply can be used with this chip.

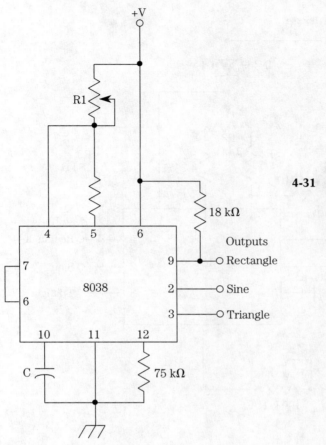

4-31 This simplified 8038 function-generator circuit has a fixed duty cycle to generate square waves, sine waves and triangle waves.

For a dual-polarity power supply, anything from ±5 volts to ±13 volts is acceptable. The usable range for single-ended supply voltages is from +10 volts to +26 volts.

When using a dual-polarity power supply, the output signals will be centered around true ground potential (0 volts). If a single-ended supply voltage is used with the XR 2206, the output signal will be symmetrical around one-half the supply voltage.

Only a bare minimum of external components are required for the XR 2206 to operate as a full-featured function generator circuit. An optional control voltage input can be used, permitting frequency modulation (FM) of the output signal, simply by inputting an ac signal as the control voltage. The XR 2206 also has built-in capabilities for *AM (amplitude modulation)*, *FSK (frequency-shift keying)* and *PSK (phase-shift keying)*. These features permit the XR 2206 to be used in a wide variety of sophisticated function generator applications.

Ignoring any input control voltage, the nominal output signal frequency generated by the XR 2206 is determined by the values of one external timing capacitor (connected between pins 5 and 6), and one timing resistor (connected between pin

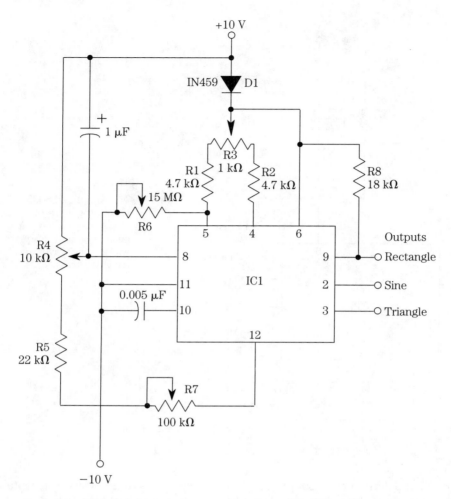

4-32   A practical audio frequency 8038 function-generator circuit.

7 or 8 and the negative supply voltage, or ground). The formula for determining the
XR 2206's nominal output frequency is certainly simple enough:

$$F = \frac{1}{R_t C_t}$$

The timing capacitor (C_t) should have a value between 1000 pF (0.000000001
farad) and 100 μF (0.001 farad). Obviously this gives you quite an extensive range to
choose from.

The timing resistor's (R_t) acceptable range of values run from 1 kΩ (1,000 ohms)
and 2 MΩ (2,000,000 ohms). However, for the best thermal stability, and to minimize
distortion of the sine wave output signal, it is best to avoid the extreme values and
restrict the value of Rt to the 4-kΩ (4,000 ohms) to 200-kΩ (200,000 ohms) range.
This is still an impressively wide range.

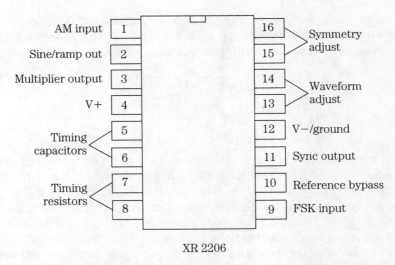

XR 2206

4-33 Another popular function-generator IC is the XR 2206.

Let's use the minimum reliable component values and find out the maximum output signal frequency for the basic XR 2206. Ct is 1000 pF, and Rt is 4Ω, so the output frequency is:

$$F = \frac{1}{(4000 \times 0.000000001)}$$

$$= \frac{1}{0.000004}$$

$$= 250,000 \text{ Hz}$$

$$= 250 \text{ kHz}$$

Notice that the XR 2206 can produce somewhat higher frequencies by decreasing the value of Rt down to 1 kΩ, but at the cost of some thermal stability and increased distortion—particularly in the sine wave output. An external control voltage can also be used to increase this maximum output signal frequency a little.

Now, let's determine the other end of the frequency spectrum, by using the largest recommended component values. In this case, Rt is 200 kΩ and Ct is 100 μF, giving us a nominal output frequency of:

$$F = \frac{1}{(200000 \times 0.0001)}$$

$$= \frac{1}{20}$$

$$= 0.05 \text{ Hz}$$

or one complete cycle every 20 seconds. Again, it is possible to get somewhat lower signal frequencies out of the XR 2206, still at the cost of some thermal stability and increased distortion.

As a more typical, mid-range example, let's use a 22-kΩ (22,000 ohms) resistor for Rt and a 0.05 µF (0.00000005 farad) capacitor for Ct. In this case, the XR 2206's output frequency (ignoring any control voltage input) will be:

$$F = \frac{1}{(22000 \times 0.00000005)}$$

$$= \frac{1}{0.0011}$$

$$= 909 \text{ Hz}$$

Earlier, it was stated that the timing resistor (Rt) could be connected to either pin #7 or pin #8 of the XR 2206. What is the point of this choice? Why didn't the designers of this chip just pick one or the other and leave it at that?

This choice gives the circuit designer using the XR 2206 more flexibility in certain specialized applications. Two separate timing resistors (with different values) can be simultaneously connected to both of these pins. Electronically switching between the two timing resistors will cause the output signal to switch between two discrete frequencies. This allows FSK (frequency-shift keying), which is a useful means of communicating and/or recording encoded data. The XR 2206 can be used as far more than just a simple function generator.

Switching between two discrete output frequencies in this manner when generating audio tones can also produce warble-tone effects. Such effects can be very useful for alarms and similar attention-getting applications. The warble-tone is harder to ignore or miss than an ordinary continuous tone. If the switching is performed at a very high rate (above about 10 to 15 Hz), the human ear will no longer be able to distinguish between the two discrete frequencies. The two tones will sonically appear to blend together into a single complex composite signal, with a lot of non-harmonic sidebands.

The active timing resistor pin (pin #7 or pin #8) is selected via pin #9. If an external voltage greater than two volts is applied to this pin, the pin #7 resistor will be active, and the pin #8 resistor will be ignored by the circuit. The pin #8 resistor is selected by feeding a voltage of less than one volt to pin #9. Usually it makes the most sense to simply ground this pin, so the input voltage is 0 volts. If pin #9 is left floating (unconnected to anything) it will internally pull itself HIGH, selecting the timing resistor at pin #7. However, floating control pins is generally not a very good circuit design practice. It is inviting the unexpected and the frustrating. I strongly recommend always connecting pin #9 to a definite HIGH (about 2 volts) or LOW (below 1 volt) point in the circuit. You can simply use the circuit's supply voltage for HIGH, or the circuit ground for LOW. It's not like you'd really save anything by leaving this pin floating.

The gain and output phase of the XR 2206's internal multiplier stage can be conveniently adjusted by applying an external control voltage to pin #1. The output is linearly controlled by variations around a nominal standard value equal to one-half the circuit's supply voltage. If pin #1 is held at this level, the output will be zero. The gain is linearly increased as the voltage on pin #1 is increased over one-half the supply voltage. Decreasing the pin #1 input voltage below one-half the supply voltage also increases the multiplier gain, but the phase is reversed.

This pin can be used for AM (amplitude modulation) and PSK (phase-shift keying) effects. For most simple function generator applications, a constant voltage is applied to this pin. Often, pin #1 will simply be grounded in a circuit using single-ended power supply. This is the maximum negative-input voltage for this pin. (Remember this signal is referenced to one half the supply-voltage). This will give us the maximum multiplier gain. The nominally reversed-phase will be irrelevant in most simple signal generation applications.

The XR 2206 is not designed to put out all of its basic waveforms simultaneously. It has just two main outputs—at pin #2 and pin #3. Pin #3 is a high-impedance output, while pin #2 is a buffered 600-ohm output. The choice will depend on the desired application and the load circuit or device to be driven by the XR 2206.

The level of the input signal to the internal buffer (the output signal at pin #2) can be varied by connecting a voltage divider between pin #3 and ground. This allows a simple method for signal gain control. This feature can also be used for keying or pulsing the pin #2 output signal.

If pins #13 and 14 are left open, the output signal will be a linear ramp or triangle wave. Placing a resistance of a few hundred ohms across pins #13 and 14 causes the peaks of the signal to be exponentially rounded off. At some specific resistance (nominally about 220 ohms), the output signal will be converted into a sine wave. With proper adjustment, the sine wave distortion can be made as low as 0.5%. For precision applications requiring a sine wave signal with minimum distortion, you can use a small-calibration trimpot between pin #13 and pin #14.

To get a true triangle wave or sine wave, the same timing resistance should be used to control both the charging and discharging cycles of the main timing capacitor. If different, widely-separated resistances are used for the two half-cycles, the triangle wave will be converted into a sawtooth wave. Many unnamed intermediate waveforms can also be generated, if desired.

A third, auxiliary output is also available at pin #11. The output signal from this pin is a rectangle wave. If a true triangle wave/sine wave is being generated at pin #2 and pin #3, the signal appearing at pin #11 will be a square wave of the same frequency. Rectangle waves with different duty cycles can be generated by choosing appropriate contrasting resistances for the charging and discharging of the main timing capacitor.

A sine-wave oscillator circuit built around the XR 2206 function-generator IC was presented in chapter 2. That same circuit can easily be adapted to a triangle-wave signal generator—just remove the resistor between pin #13 and pin #14, as illustrated in Fig. 4-34. This circuit was discussed in chapter 2, so there would be little point in repeating that information here.

Figure 4-35 shows a square-wave signal generator circuit built around the XR 2206. A load/pull-up resistor is connected to the sync output (pin #11) and the output signal is tapped off from this connection. A signal taken from this output can only be used with high-impedance loads. This output cannot be used to drive a low-impedance load, such as a standard 600-ohm audio line. If you must drive a low-impedance load from this circuit, you must use an external buffer amplifier stage. An op amp voltage follower usually works quite well, and is very simple and inexpensive to build.

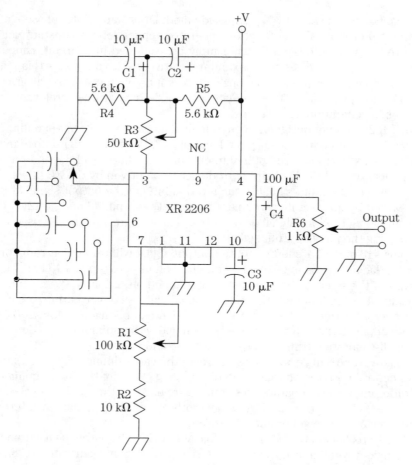

4-34 A simple but practical triangle-wave signal-generator circuit built around the XR 2206 function-generator IC.

Project #5—
Three-waveform function generator

Dedicated function-generator ICs like the 8038 and XR 2206 are great for many applications, but sometimes they can be rather expensive, and even a little tricky to find. For many applications, their superior specifications are overkill. If you are interested in function generator projects built around these chips, the sample circuits described throughout this chapter should be sufficient to get you started.

For our project in this chapter, however, we will use a simpler, discrete circuit using three transistors and no ICs at all. This project is quite inexpensive, easy to construct, and easy to work with. The requirements for the three transistors are flexible enough that many standard substitutes can be used with no problem. The schematic diagram for our function generator project is shown in Fig. 4-36. A suitable parts list for this circuit is given in Table 4-2.

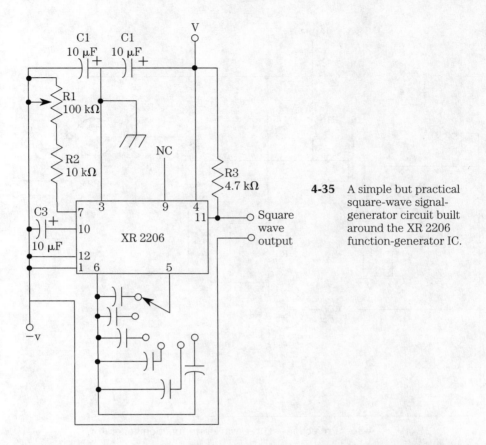

4-35 A simple but practical square-wave signal-generator circuit built around the XR 2206 function-generator IC.

This function generator project can generate any of three standard wave-forms. Only one waveform can be fed to the output at once. Changing the output waveform does not affect the output signal's frequency. Rotary switch S2 selects the waveform. When this switch is set to position D, there will be no output signal at all. This dead position is included only because it will probably be easier to find a SP4T rotary switch than a SP3T unit.

The three waveforms produced by this circuit are:

    A   Triangle wave
    B   Square wave
    C   Sawtooth wave
    D   Dead position—no output signal

This circuit does not generate a sine wave. In many applications, a triangle wave is an acceptable substitute for a sine wave, due to the fact that its harmonic content is so weak. For even better results, try passing the triangle wave output signal through a low-pass filter to get a better approximation of a sine wave (even less harmonic content).

For some applications, it might be a limitation that only one of the three generated waveforms can be used at any given moment. In most practical applications,

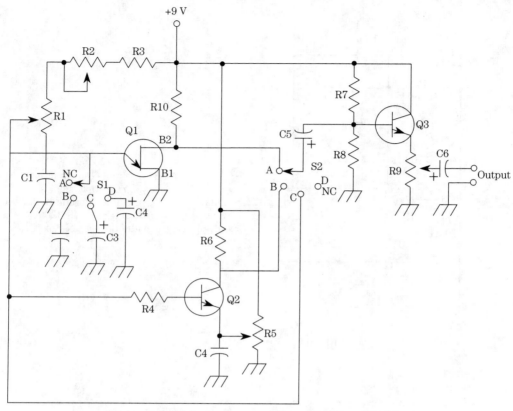

4-36 Project #5—Three-waveform function generator.

**Table 4-2. Suggested parts list for Project #5—
Three-waveform function generator of Fig. 4-36**

Q1	UJT (2N2646, Radio Shack RS2029, or similar)
Q2, Q3	pnp transistor (2N4124, GE-20, or similar)
C1	0.01-µF capacitor
C2	0.1-µF capacitor
C3	1-µF, 35-V electrolytic capacitor
C4	10-µF, 35-V electrolytic capacitor
C5, C6	47-µF, 35-V electrolytic capacitor
R1	25-kΩ potentiometer (frequency coarse tune)
R2	500-Ω potentiometer (frequency fine tune)
R3	2.2-kΩ, ¼-W, 5% resistor
R4, R6	10-kΩ, ¼-W, 5% resistor
R5	50-kΩ trimpot (distortion adjust)
R7, R8	100-kΩ ¼-W, 5% resistor
R9	5-kΩ potentiometer (signal amplitude adjust)
R10	1.5-kΩ, ¼-W, 5% resistor
S1, S2	SP4T (single-pole, four-throw) rotary switch

you will probably need only a single waveform at a time anyway, so this will rarely be a significant disadvantage, and it simplifies the circuit design considerably.

This function generator project is manually adjustable over a fairly wide range of frequencies. It can generate most audible frequencies, and even some sub-audio frequencies, all at minimal expense and circuit complexity.

Switch S1 serves as a frequency range control for the project. It selects one of four timing capacitors, setting up four overlapping frequency ranges. Actually, capacitor C1 is always in the circuit, regardless of the position of the switch, so it is effectively in parallel with one of the other timing capacitors when the switch is moved past its first (no connection) position. Capacitances in parallel add, but because the value of capacitor C1 is so small, it won't change the total capacitance significantly. Ignoring normal component tolerances, the nominal and actual capacitance values for each switch position are as follows:

Switch position	Nominal capacitance	Actual capacitance
A	0.01 μF	0.01 μF
B	0.10 μF	0.11 μF
C	1.0 μF	1.01 μF
D	10.0 μF	10.01 μF

Obviously, the normal tolerance rating of most practical capacitors would account for more error than this.

Two potentiometers, R1 and R2, are used to tune the actual frequency within the specified range. R1 is a coarse tuning control, and R2 is used for precise fine tuning. You can probably get away without R2 in many non-critical applications, but R1's range is so large that it might be difficult to set it precisely. The fine-tuning control can make the function generator a lot easier to work with.

Nominally, you would need only one calibration dial on the coarse tuning potentiometer (R1). On other ranges, you'd just have to mentally move the decimal point. Each range changes the frequency by a factor of ten. In practice, there will probably be some additional fluctuation, due to the tolerances of the specific capacitors used. In most cases, it will just be a matter of re-adjusting the fine-tuning control. This control should not be calibrated with frequency values, because they will be totally inaccurate once potentiometer R1 is moved even slightly.

Figure 4-37 illustrates one good way to set up the front panel of your function generator project, and mark the controls. The dial around potentiometer R1 can be calibrated by using a frequency counter or oscilloscope to monitor the output signal as this control is moved through its range. Using the component values suggested in the parts list, the four frequency ranges overlap each other by quite a bit, so no gaps occur in the total range of available output frequencies.

Depending on the accuracy (tolerance) of the components you use in your project, the overall frequency range for this project should pretty much cover the entire audible spectrum from about 20 Hz up to 20 kHz. This is assuming that you use the component values suggested in the parts list, of course. You should be aware that this simple circuit can get a little fussy and unreliable when it is forced to generate frequencies below about 200 Hz. You might have to do a little extra fiddling with the controls when switch S1 is set to frequency range D.

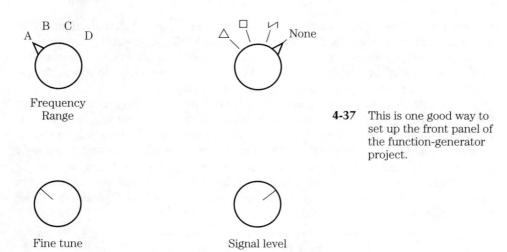

Frequency
Range

Fine tune

Signal level

4-37 This is one good way to set up the front panel of the function-generator project.

The specifications for the transistors used in this circuit are not critical. If you can't find any of the specific types suggested in the parts list, check a good transistor substitution guide. Almost anything listed as similar should work just fine in this project. When in doubt, breadboard the circuit and test it thoroughly before you do any soldering.

There are two other controls in this circuit that we haven't considered yet. Potentiometer R9 is used as an output amplitude control, or volume control when you use the function generator to produce audible sounds. This control permits you to adjust the output signal level to suit the particular load circuit or device you are driving with the function generator. In some applications where a variable output amplitude is not needed, this potentiometer could be replaced with a fixed resistor.

R5 should be a screwdriver-adjust trimpot, not a front panel control potentiometer. Once properly calibrated, this trimpot should be left alone. While monitoring the output signal with an oscilloscope, slowly and carefully adjust trimpot R5 for the cleanest possible waveforms. Check the signal purity at several different frequencies in each frequency range selectable via switch S1. Also, check all three output waveforms (selected by switch S2). Sometimes changing the signal frequency or the waveshape will require some additional readjustment of R5. In some cases, you might have to accept a compromise calibration. It can be a nuisance to perform this picky calibration to the needed extent, but once it is done, you shouldn't have to worry about it again—especially if you put a small drop of paint or glue on R5 to hold its wiper in place so that it can't physically readjust itself as the project is moved about.

Project #6—Sweep-signal generator

Does this sweep-signal generator project really belong in this chapter on function generators? Perhaps not, but, frankly, I didn't know where else to put it, and I felt it was too important and useful of a project to leave out. Yet, it didn't seem to warrant its own separate chapter. So, for better or worse, it seemed to fit at least as well here as anyplace else in this book.

Most signal generators are designed to put out just a single frequency at a time. In many circuits, the frequency can be manually adjusted, but once set, it tends to stay put (ignoring any gradual random drift effects), until it is physically re-adjusted. In most applications, this is exactly what we want. But in some specialized applications—especially on the electronics test bench—it can be helpful, if not essential, to have a signal generator that can automatically slide through its frequency range. This is very useful for testing the frequency response and/or stability of an amplifier or other circuit, among other applications. This type of semi-automated equipment is called a *sweep-signal generator*. The generated output signal is swept through its frequency range.

Most practical sweep-signal generator circuits use some sort of *VCO (voltage-controlled oscillator)* to generate the output signal. The frequency of this output signal is proportional to a control voltage signal at the VCO's input. In a sweep-signal generator, the control voltage signal is a low-frequency sawtooth wave, like the one shown again in Fig. 4-38.

4-38 In a sweep-signal generator, the control voltage signal is a low-frequency sawtooth wave.

Notice how the instantaneous voltage of this waveform changes over time. It starts out at its minimum (most negative) value, then smoothly and linearly increases up to its maximum (most positive) value. The signal voltage then drops very quickly back (theoretically there is no transition time) to the minimum level again, and the cycle repeats itself.

The VCO, being driven by this sawtooth-wave control voltage, automatically adjusts its output frequency so that it is proportional to the instantaneous control voltage presently at its input. Therefore, it beings each cycle at its lowest output frequency, then smoothly and linearly glides up to its highest output frequency, then drops back down to the lowest end of the frequency range to begin the next cycle. The signal sweeps through its entire frequency range from lowest to highest, in a very predictable and repeatable way. The control voltage sawtooth wave has a very low frequency, well below the audible range. But we can hear the gradual changes in signal frequency, or observe them on an oscilloscope, or on some other piece of test equipment.

By monitoring the output of an amplifier (or other circuit) being fed by this swept signal, we can easily spot any sudden changes in the output signal's amplitude, indicating an uneven frequency response in the circuit or device being tested. With an oscilloscope, we can also determine the approximate frequency where the signal starts to become distorted.

Yes, similar tests could probably be done by manually adjusting the frequency of a standard signal generator or function generator through its range, but this would take a lot more time and would be a lot of unnecessary and inconvenient fuss and bother. The sweep-signal generator does some of the mindless busy work for us, so we can concentrate on observing the output signal during the test procedures.

Commercial sweep-signal generators are often very sophisticated and expensive, with a lot of special features that are of little use to the average electronics hobbyist. In this project, we will accomplish the same basic thing at a much lower cost. A typical commercial sweep-signal generator sells for at least $200 or $300. You should be able to build this circuit for $10 to $20, and maybe even less if you find a good source of surplus components, or have a well-stocked electronics junk-box.

No, the specifications of this inexpensive home-made unit aren't nearly as good as most professional sweep-signal generators, but they should still be good enough for all but the most critical of hobbyist applications. For most purposes, the better specifications of the commercial equipment would just be expensive over-kill and wouldn't give you any real practical advantage.

The complete schematic diagram for this simple, inexpensive-but-powerful sweep-signal generator project is shown in Fig. 4-39. A suitable parts list for this project is given in Table 4-3.

For convenience in our discussion, we will divide the circuitry into two functional sections, as shown in the block diagram of Fig. 4-40. The first section is the sawtooth-wave generator (control voltage source), made up of IC1, IC2, IC3, and their associated components. The second section is the VCO, which generates the actual output signal under the control of the first stage. This VCO is made up of IC4 and its associated components.

Looking at only the first circuit section for now, this is just a low-frequency sawtooth-wave oscillator or signal generator circuit. In fact, if you just want to build a sawtooth-wave signal generator project, you can use this circuit—just select appropriate values for the frequency-determining components (resistor R7 and capacitor C1) and eliminate the following components:

IC4
C2
R10
R11
R12

Even in the complete sweep-signal generator project, you can make a provision for tapping off only the straight sawtooth-wave signal directly from the output of IC3.

All three of the ICs in the sawtooth-wave generator section of this circuit are simple op amps. Almost any op amp devices can be used, even inexpensive 741s should work fine in this project. I recommend using three sections of a 324 quad op amp IC in this project. The fourth section of this chip can be used for some other circuitry, or just left unused—it won't do any harm. The advantage of using the 324 is that, unlike the standard 741, it can be easily used with a single-polarity power supply. The 741 op amp is designed for use with a dual-polarity power supply. Some newer, better, single op amp ICs can be operated by a single-ended supply voltage. Check the manufacturer's specification sheet for the power supply requirements for the specific op amp IC(s) you decide to use in your project.

Notice that no power supply connections are shown to the op amps in the schematic diagram. This is because there are so many possibilities, depending on whether you are using a single- or dual-polarity power supply, and whether you are

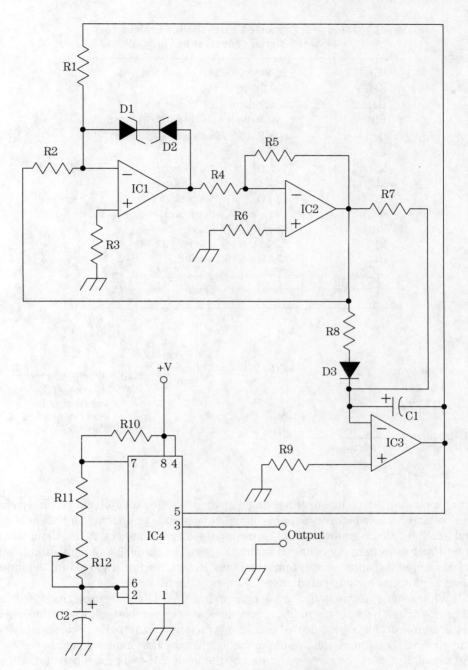

4-39   Project #6—Sweep-signal generator.

**Table 4-3. Suggested parts list for Project #6
—Sweep-signal generator of Fig. 4-39**

IC1, IC2, IC3	Op amp (see text)
IC4	555 timer
D1, D2	Zener diode (5.1 V—see text)
D3	Signal diode (1N4148, 1N914, or similar)
C1	50-μF, 35-V electrolytic capacitor*
C2	0.01-μF capacitor**
R1, R2, R4, R5	10-kΩ, 1/4-W, 5% resistor
R3, R6	4.7-kΩ, 1/4-W, 5% resistor
R7	68-kΩ, 1/4-W, 5% resistor*
R8, R11	1-kΩ, 1/4-W, 5% resistor
R9	62-kΩ, 1/4-W, 5% resistor
R10	2.2-kΩ, 1/4-W, 5% resistor
R12	25-kΩ potentiometer**

*Experiment with alternate values to change sweep frequency
**Experiment with alternate values to change signal frequency

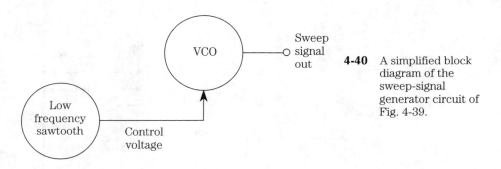

4-40 A simplified block diagram of the sweep-signal generator circuit of Fig. 4-39.

using a quad-op amp chip, or three single op amp ICs, or a dual-op amp IC and a single-op amp IC, or whatever. Supply voltage connections to op amps are often omitted in schematic diagrams to avoid cluttering the diagram unnecessarily. Remember, the proper power supply connections must always be made. No op amp circuit will work without the appropriate supply voltages. In fact, feeding a signal into the input of an unpowered op amp can damage or destroy the IC.

While even a relatively low-grade op amp like the 741 can be used (its specifications are actually pretty good, but newer devices are even better), if your intended application calls for a great deal of precision, you might want to use better-quality op amp devices. Low-noise units would probably offer the best improvement in the output signal. But even the noise generated by the lowly 741 will probably be unnoticeable in most practical applications.

IC1 is wired as a linear integrator. IC2 and IC3 form a feedback loop to cause different integration rates for the ascending (sweep) and descending (drop-back) portions of the cycle. To generate a good sawtooth wave signal, we want the ascending portion of the waveform to last a relatively long time, and the descending portion to last as short a time as possible. Therefore, resistor R8, which is functionally part of the circuit only when diode D3 is forward-biased, has a very small value, compared to that of resistor R7. Therefore, the descending portion of the cycle is much, much faster than the ascending portion. No, it isn't a truly instantaneous drop-back, but it is very fast, and should be good enough for most practical applications. If you would prefer to make the circuit generate a triangle wave all you have to do is make resistor R8 equal to R7.

In the sawtooth-wave generator circuit, the drop-back time is so short that it can be reasonably ignored. Therefore, the signal frequency is determined solely by the time constant of the ascending ramp. The approximate frequency of the sawtooth-wave generated by this part of the circuit can be found with this formula:

$$F = \frac{1}{(R_7 C_1)}$$

F is the signal frequency in hertz. R_7 is the resistance in ohms, and C_1 is in farads.

Notice that this formula is only approximate, because of normal component tolerances, and the fact that the drop-back time of each cycle is being ignored. Usually the exact frequency of the sweep cycle isn't very critical anyway.

Using the component values suggested in the parts list, the sweep signal will have an approximate frequency of:

$$F = \frac{1}{(68,000 \times 0.00005)}$$

$$= \frac{1}{3.4}$$

$$= 0.3 \text{ Hz}$$

Each complete sawtooth wave cycle produced by this circuit will last about 3.5 seconds.

You can change this sawtooth-wave sweep frequency by changing the value of either capacitor C1 or resistor R7, or both. Increasing either or both of these component values will lower the sweep signal's frequency proportionately.

In some applications, you might want to have a manually-adjustable sweep rate. That is easy enough to implement. Just replace resistor R7 with a 100-kΩ potentiometer in series with a 10-kΩ resistor. If the capacitor value is unchanged (50 µF), the manually-adjustable sweep frequency range will run from 0.18 Hz (one cycle every 5.5 seconds) up to 2 Hz (two complete cycles a second). Other component values will give different sweep rates.

Zener diodes D1 and D2 set the minimum and maximum control voltages fed to the VCO, respectively. Using 5.1-volt zener diodes, as suggested in the parts list, results in a 10-volt peak-to-peak sawtooth wave signal. The control voltage will range from –5.1 volts to +5.1 volts. This peak-to-peak output voltage can be altered by using different zener diodes, or a resistive voltage-divider network can be placed across the output of IC3. In the later case, it would not be advisable to try to tap off the straight sawtooth-wave signal without adding a buffer amplifier stage, or loading effects might be extreme.

Incidentally, if it suits your application, you can also tap off a secondary low-frequency output signal at the output of IC2. The waveform at this point in the circuit is string of very narrow positive-going spikes, at the same frequency as the sawtooth wave signal generated at the output of IC3.

Now, let's turn our attention to the second half of this sweep-signal generator—the VCO (IC4 and its associated components). IC4 is just a common 555 timer chip, wired as a standard astable multivibrator circuit with a control voltage signal (from IC3) fed into pin #5. Because this control voltage input is being actively used here, it should not be bypassed to ground with the usual stability capacitor recommended in most 555 based circuits.

Of course, the output waveform is a rectangle wave. The duty cycle is determined by the ratio of resistor values, $R_{10}:(2R_{10} + R_{11} + R_{12})$. Changing the position of the manual frequency control (potentiometer R12) will also affect the duty cycle, but the duty cycle will not be affected by changes in the output signal frequency due to the changing control voltage.

Whenever the control voltage signal has an instantaneous value other than zero, the output signal frequency will be deflected proportionately from its nominal value (set by the component values in this portion of the circuit). Potentiometer R12 permits manual adjustment of the base output frequency, and thus the frequency range covered by the swept signal. If your intended application doesn't need a manually-adjustable base frequency, you can combine resistor R11 and potentiometer R12 into a single fixed resistor of a suitable value.

The usual 555 astable multivibrator frequency equation applies to this circuit. The nominal base signal frequency (ignoring the sweeping control voltage) is equal to about:

$$F = \frac{1.44}{\{C_2 \times [(R_{11} + R_{12}) + (2 \times R_{10})]\}}$$

Setting the potentiometer to the mid-point of its range (12.5 kΩ or 12,500 ohms), and using the component values suggested in the parts list, the nominal base output signal frequency for this circuit works out to approximately:

$$F = \frac{1.44}{\{0.0000001 \times [(1000 + 12500) + (2 \times 2200)]\}}$$

$$= \frac{1.44}{[0.0000001 \times (13500 + 4400)]}$$

$$= \frac{1.44}{(0.0000001 \times 17900)}$$

$$= \frac{1.44}{0.00179}$$

$$= 805 \text{ Hz}$$

The sweeping control voltage from the first half of the circuit causes the actual output frequency to drop well below and rise well above this nominal base frequency as it sweeps through its range. This nominal base frequency occurs at the exact mid-point of the sawtooth-wave sweep signal.

5
CHAPTER

Pink-noise generators

A noise generator might sound like an utterly ridiculous and useless idea. After all, noise is always something undesirable and to be avoided, isn't it? Why on Earth would anyone want to deliberately generate it?

Like many, many words in the English language, *noise* has different meanings in different contexts. In general usage, noise usually means an unpleasant or annoying sound, that is probably rather loud. Banging on a metal garbage can is noise, while playing a musical instrument isn't (unless you happen to really dislike the music being played, or it is being played very badly).

In electronics, the word noise usually refers to an undesired signal in a circuit or a system. The word is less judgmental here. There might not be anything unpleasant about the noise itself, it simply isn't the signal we want. For example, when you tune a radio to a rock music station, and you get interference from a classical music station, a symphony can be noise. Or the rock station might interfere with the classical music station you want to listen to, then the rock music is noise. You might be a person with wide tastes, like me, and enjoy both rock music and classical music—but when you want to listen to one, interference from the other is undesired, making it noise.

Except for undesired RF pick-up from radio stations, most of the undesired signals called "noise" are more like noise in the usual, acoustic sense (unpleasant sound). (Actually, electronic noise doesn't have to be a sound at all—it can be an undesired control signal, or a signal affecting the reading of a meter, or any other electronic function.) For example, there can be electromagnetic pick-up of 60 Hz hum, or a transistor or op amp, or another active device might internally generate a small, randomized signal that could be amplified by the circuit until it reaches an annoyingly undesirable level.

But this suggests yet another way we can define "noise." It is a random (or pseudo-random) signal, rather than a neat repeating cycle, such as a sine wave or a square wave. On an oscilloscope, noise (in this sense) just looks like a mess, as illustrated in Fig. 5-1.

5-1 On an oscilloscope screen, noise looks like a jumbled mess.

White noise

Ordinarily, it seems rather odd to say noise has a color. After all, noise is a sound, and color is a visual phenomena. Still, it is a useful analogy for certain technical work.

Light, like sound, can exist at any of a wide range of frequencies. Only a very narrow band of specific frequencies can be seen directly with the naked eye. This band of frequencies is called the *visual spectrum*. Within the visual spectrum, a change in frequency is perceived as a change in color. The lowest visual frequencies look red, and the highest visible frequencies look purple. All other colors have frequencies somewhere in between these extremes.

Most natural light sources, including the sun and most light bulbs, emit a wide variety of light frequencies, all jumbled up together. The range of light frequencies put out by a standard light bulb extends past the visual spectrum in both directions. In other words, the entire visual spectrum is included in the light bulb's output. The sun puts out an even wider range of light frequencies—including the entire visual spectrum.

Every specific frequency in the light source's range is equally likely to be emitted at any given instant. In other words, over time, there is equal energy at each specific frequency within the visual spectrum (which is all we are interested in for visible light). How does this look? It is white light. White is the color perceived when all visible light frequencies are mixed together in approximately equal proportions. In electronics, *white noise* is made up of equal energy at each specific frequency within the audible range (roughly from 20 Hz up to 20 kHz [20,000 Hz]).

If you tune an FM radio without muting between stations, the hiss you will hear is white noise. Notice that it doesn't really sound like it's made up equally of all audible frequencies. Low frequencies (pitches) seem to be short-changed. It sounds like a high-pitched hiss. This is because of the way the human ear hears various frequencies.

The reason for the perceived high-pitch bias of white noise is contained within the acoustic concept of octaves. An octave is a doubling of frequency. If we start out with a 100 Hz, raising it one octave, we get a 200-Hz signal. Raising that same 100-Hz signal two octaves results in a 400-Hz signal. Because of the internal design of the human ear, two musical tones separated by an octave, will sound like the same note, even though the difference in pitch is obvious. One is much higher in pitch than the other, yet they still have a sort of unity between them. They sound perfectly in tune with one another. This is why it usually is more reasonable to speak of audible frequency ranges in terms of octaves rather than absolute frequency differences. As the frequency increases, the octaves get larger in absolute terms, even though they sound equivalent. This is illustrated in the graph of Fig. 5-2.

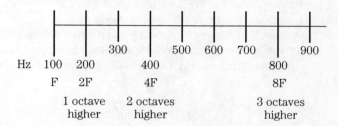

5-2   As the frequency increases, octaves get larger in absolute terms, even though they sound equivalent.

A high octave, say 6.4 kHz (6,400 Hz) to 12.8 kHz (12,800 Hz), is made up of more discrete frequencies than a lower octave, say 200 Hz to 400 Hz. The low octave from 200 Hz to 400 Hz covers a range of just 200 Hz. But the higher octave in our example has a range extending 6,400 Hz. Yet both octaves are perceived as equal by the human ear.

Take another look at the frequency distribution graph of Fig. 5-2. It should be apparent from this graph that a white noise signal with equal energy-per-frequency is going to inevitably make the higher octaves sound "top-heavy." There are more discrete frequencies-per-octave at the upper ranges.

The white noise between FM stations sounds high-pitched simply because there are statistically more high-pitched frequencies (as the ear perceives them) than low-pitched frequencies. The human ear just doesn't hear frequencies linearly, so a linearly equal mix of all audible frequencies won't sound equal. Another graph illustrating this is shown in Fig. 5-3.

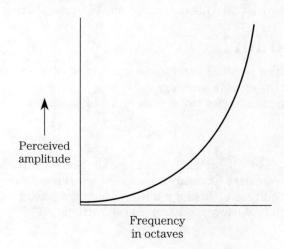

5-3   The human ear doesn't hear frequencies linearly, so a linearly equal mix of all audible frequencies won't sound equal.

Pink noise

To achieve a noise signal that sounds evenly distributed, we need to use pink noise rather than white noise. Again, the color reference is a handy analogy with light. Remember, the lowest visible frequencies are seen by the eye as red. If we mix red with white, we get pink. We've already seen that white light contains equal energy-per-visual-frequency. Mixing in the red light means that the overall energy content of the lower frequencies will be relatively boosted.

Similarly, in pink noise, the lower octaves are given extra emphasis, so it sounds like the audible spectrum is evenly balanced. White noise has equal energy-per-frequency, but pink noise has equal energy-per-octave. The two "colors" of noise are compared in Fig. 5-4.

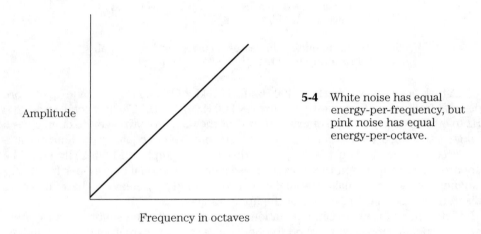

Amplitude

Frequency in octaves

5-4 White noise has equal energy-per-frequency, but pink noise has equal energy-per-octave.

In most circuits, the easiest and most obvious way to create pink noise is to start out with white noise, then pass this signal through a low-pass filter, which passes low frequencies with less attenuation than high frequencies. Some pink-noise generators produce their equal energy-per-octave signal without needing any low-pass filtering.

What is noise good for?

OK. So now we know what white noise and pink noise are, but what are they good for? Why would anyone want to generate such signals?

The three primary classes of application for white noise and pink noise are:
- Sound effect synthesis
- Broad-band testing
- Random-voltage generation

Almost any sound (real or imagined) theoretically can be synthesized electronically. (Some naturally occurring [acoustic] sounds are impractical to synthesize accurately because of limitations of current technology, and the high complexity of certain sounds—especially the sounds of many acoustic musical instruments.)

In most cases, the primary signal source is an audio-frequency oscillator circuit of some sort. If we start out with an oscillator tone, the final synthesized sound will probably have a strong sense of pitch. But many sounds we might want to synthesize are more or less unpitched, especially when we are dealing with sound effects rather than musical notes. Typical unpitched sounds include those of drums, rain falling, thunder, explosions, gunshots, footsteps, and many others. A pink-noise or white-noise signal is a good starting point for synthesizing such unpitched sound effects. The noise signal can be filtered and enveloped to create the desired effect.

Sound synthesis is not just used in musical synthesizers. Other applications include computer games, toys, and theatrical sound effects, among others.

Pink noise and white noise can be very useful as test signals for audio amplifiers and similar equipment. By definition, these signals cover the entire audible frequency spectrum at once. Any severe fluctuations in the frequency response of the amplifier (or whatever) being tested can be made immediately apparent, just by listening to the sound at the output speakers when a pink-noise or white-noise generator is used as the input signal.

Because few practical amplifier circuits or (especially) speakers give a truly flat-frequency response, the amount of gain can vary from signal frequency to signal frequency. By using pink noise or white noise as the test signal, we can easily measure the overall gain covering the entire audible frequency spectrum.

A handy "quick and dirty" test of an amplifier's output power can be made with an ordinary light bulb. Unlike most such tests, there is no need for expensive test equipment or calibrated resistors in the test procedure. An ordinary single-frequency signal generator can be used as the input signal in such a test, but the results will only be accurate for the specific test frequency. You will usually need to repeat the test several times at various signal frequencies to be sure of the results. If a white-noise or pink-noise generator is used as the signal source, however, the entire audible frequency range will be covered at once, and you can reliably check out the amplifier's average output power with a single quick test.

This test is only advised for a PA amplifier, or something similar, with a high-impedance output. For a standard audio amplifier, such as a stereo, use a separate impedance-matching transformer. This transformer should be enclosed in a well-insulated case for safety's sake.

To set up this test, get a weatherproof lamp socket and put a couple of heavy terminal lugs on the wires—spade lugs are best. Connect this socket across the high-impedance output of the amplifier (or to an impedance-matching transformer). Connect a pink-noise or white-noise generator to the amplifier's input.

Before powering up the amplifier, screw a standard incandescent light bulb into the socket. This light bulb should match the amplifier's nominal output power wattage as closely as possible. There is some reasonable leeway. For a 50-watt amplifier, use a 50-watt lamp. For a 30-watt amplifier, you should probably use a 25-watt lamp (the nearest standard size). This simple test set-up is illustrated in Fig. 5-5.

Now power up the amplifier, and watch the light bulb light up. If you have the rated power output from the amplifier, the lamp should light up with full brilliance. If it does, everything is fine, and the test procedure is complete.

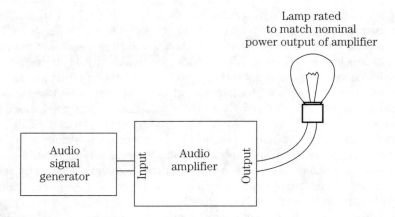

5-5 A common light bulb can be used to roughly test an amplifier's output power.

If the lamp doesn't light up at all, something is wrong. Double-check the test wiring. Make sure the light bulb you are using is good and not burnt out. Make sure the noise generator is putting out a signal. Make sure that all of the amplifier's controls are set correctly. If all of these things are correct and the lamp still doesn't light up, the amplifier is defective. You will need more detailed and precise testing to service it.

If the lamp does glow, but only dimly, the amplifier's power output is too low for some reason. Again, further, more detailed troubleshooting is called for.

What happens if you choose the wrong wattage for the light bulb? Nothing too terrible, really. Mostly, you just won't get very accurate results from the test. If the lamp's wattage is significantly higher than the amplifier's output power, it will not be able to glow with full brilliance, even if the amplifier is working properly. On the other hand, if the lamp's wattage is too low for the tested amplifier's output power, it will probably burn out. No damage will be done, except to the light bulb itself.

This crude test method is generally not recommended for high-power transistor amplifiers because of their more critical impedance-matching requirements. In such cases, if you do try this type of test, an appropriate impedance-matching transformer is essential, or else you could seriously damage the output stage of the amplifier you are trying to test.

Obviously, that lamp test is pretty crude and inexact. It's a handy trick for certain emergency situations. But in many practical applications, a more exact indication of an audio amplifier's output power is needed. Whatever method is used, an input signal from a pink-noise or white-noise generator is a good test signal because it averages out the amplifier's gain over the entire audible frequency spectrum.

The most obvious way to measure the output power of an amplifier is to use a wattmeter—a piece of test equipment designed specifically for that purpose. Unfortunately, such a device is not a standard item on most electronics workbenches. If you do a lot of work on audio amplifiers, a wattmeter would probably be a good investment, but for most general electronics hobbyists and technicians, it will probably spend most of its time just gathering dust.

Luckily, you don't absolutely need a wattmeter to measure the output power of an audio amplifier with reasonable accuracy. You can make such a test with a suitable load resistor and the ac voltmeter section of a multimeter. Either an analog or digital multimeter can be used equally well in this test.

The basic test set-up is illustrated in Fig. 5-6. Notice that a load resistor is used in place of the normal loudspeaker. This is to avoid the complexities and probable misreadings that are likely to result from the ac impedance of the speaker's voice coil. A noise signal would help to average out these effects, but you still probably won't get an accurate output wattage reading. Everything will be much simpler and straightforward if only a straight dc resistance (a simple load resistor) is involved.

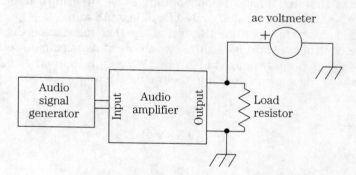

5-6 This is an improved method of testing an amplifier's output power without an actual wattmeter.

The load resistor's value should be equal to the nominal impedance of the speaker it is temporarily replacing. For most modern stereo amplifiers, this will probably be 8 ohms. Other common values you might encounter are 4 ohms and 16 ohms. Some PA amplifiers might have even higher output impedances. In most cases, the correct output impedance value is indicated on the back of the amplifier near the output jack or screw terminals. The resistor value does not have to be perfectly exact. Remember, the ac impedance of a loudspeaker will vary with the signal frequency, so there is some leeway in what the amplifier's output is designed to accept. For example, if you need a 16-ohm load, a standard 15-ohm resistor should be close enough in most cases.

More importantly, a high-power resistor must be used in this application. The load resistor's power rating should at least be equal to the expected power output of the amplifier being tested. That is, if you are testing an amplifier that is supposed to put out 30 watts, use a load resistor rated for at least 30 watts, and preferably more. It is a good idea to over-rate the load resistor's power rating by at least 10% to 25%. To test an amplifier with a nominal output power of 30 watts, the load resistor should be at least 33 watts to 37.5 watts. If it is larger than this, that's even better.

If the amplifier being tested has any tone controls or filters, set them all for flat response while performing this test procedure.

You can use the ac voltage measured across the load resistor to determine the output power with this standard formula:

$$P = \frac{E^2}{R}$$

$$= \frac{(E \times E)}{R}$$

where P is the power in watts, E is the measured voltage in volts, and R is the load resistance in ohms.

As an example, let's assume we are testing an amplifier with a nominal output power rating of 15 watts. It has a 16-ohm output, so we are using a 15-ohm load resistor (the closest convenient value), rated for at least 18 watts.

It is a good idea to first perform this test at less than full volume. Set the amplifier's volume control at about one-half to two-thirds of its maximum setting. Now, let's say we read an output voltage of 9.5 volts. What is the output wattage? All we have to do is plug the known values into the standard power equation:

$$P = \frac{9.5^2}{15}$$

$$= \frac{(9.5 \times 9.5)}{15}$$

$$= \frac{90.25}{15}$$

$$= 6.0 \text{ watts}$$

This seems to be a fairly reasonable half-volume output wattage for a 15-watt amplifier. Now, let's carefully advance the amplifier's volume control to its maximum setting. If the load resistor gets too hot, you are putting too much power across it. It's not advisable to try to touch it, because it is carrying live ac power, which could result in a painful or even fatal electrical shock. But if the air a few inches away from the load resistor feels abnormally warm, or if the resistor starts to change color or glow, or if you smell smoke, immediately turn off the amplifier's power switch. The over-stressed load resistor will change value and could conceivably damage the amplifier's output stage if left running for more than a couple of seconds. If you properly selected the power rating of the load resistor with plenty of over-rating, this is not likely to be a problem.

Now, to return to our example, lets say that at full volume, this amplifier has a measured output voltage of 14.75 volts. This means that its maximum power output is:

$$P = \frac{14.75^2}{15}$$

$$= \frac{(14.75 \times 14.75)}{15}$$

$$= \frac{217.5625}{15}$$

$$= 14.5 \text{ watts}$$

This is a little low from our nominal value, but only by half a watt. There might be a half-watt of error in the normal tolerances of the test procedure. This amplifier should be considered close enough to its rated specifications. Similarly, if the measured output power comes out a little too high, that is not a problem. The manufacturer probably rounded off the actual output power specification. A 16.5-watt amplifier would probably be rated for only 15 watts to provide some "headroom," and to give a standard value that will be easier for the sales literature for the amplifier.

On the other hand, suppose we checked out a second amplifier with the same ratings as the first, but this time at full power we measure an output voltage of 11.0 volts. This would mean the maximum power output is:

$$P = \frac{11^2}{15}$$

$$= \frac{(11 \times 11)}{15}$$

$$= \frac{121}{15}$$

$$= 8.1 \text{ watts}$$

This is way too low, and not close enough to the rated power output specification (15 watts). Further troubleshooting of this amplifier is called for.

This test is quite useful and reasonably accurate, but it is still rather crude. It doesn't really give you enough information to fully judge the amplifier. You know the absolute output signal amplitude, but you don't know how badly the signal is being distorted by the amplifier, or how smooth the frequency response is. Other tests will be needed to identify such problems. An oscilloscope can be used to spot distortion problems. An oscilloscope with a sweep-signal generator (see chapter 4) can be used to check out the amplifier's frequency response. Other tests using pink noise or white noise as input signals are also possible, and are not uncommon in practical electronics work.

We will consider the third class of applications for pink noise and white noise—generating random, or pseudo-random voltages—a little later in this chapter. First, we will take a look at a few practical noise generator circuits.

Noise-generator circuits

It is not very difficult to generate pink noise or white noise electronically. After all, it happens to some extent in almost every signal-carrying circuit anyway, whether we want it or not. In most applications, the circuit is designed to minimize noise generation as much as possible, but in a pink-noise generator or white-noise generator, we take advantage of this natural and inevitable tendency, and deliberately try to maximize it.

As you might suspect, noise generators tend to be rather simple circuits. They are rarely expensive or bulky. They usually don't have many special features. Often a noise generator will have no controls at all, except for a power switch to turn it on and off. Some noise generators have an output amplitude (volume) control, and maybe an extra switch to select between pink noise or white noise, but that's about it. Simplicity and straightforwardness is by far the norm for this type of circuit.

Traditionally, most analog noise generator circuits are built around a natural defect of all practical semiconductor components. Virtually all practical diodes and transistors generate a certain amount of noise whenever current flows through them. This internally-generated noise signal is typically very small (low in amplitude), but it is easy enough to amplify it up to a usable level.

Occasionally, you might build a noise generator circuit that won't produce a steady stream of noise. If this happens, simply try another diode or transistor of the same type number. Every now and then you'll come across one that isn't particularly noisy, so that individual unit won't work in this admittedly unusual application.

A fairly typical and exceptionally simple noise-generator circuit built around a diode is shown in Fig. 5-7. A suitable parts list for this simple circuit is given in Table 5-1.

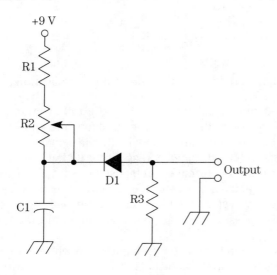

5-7 This is a fairly typical and exceptionally simple diode-based noise-generator circuit.

Table 5-1. Typical parts list for the simple diode-based noise -generator circuit of Fig. 5-7

D1	Almost any diode (Radio Shack 276-1101, or similar)
C1	500-pF capacitor
R1	470-Ω, ¼-W, 5% resistor
R2	100-kΩ potentiometer
R3	47-Ω, ¼-W, 5% resistor

A fixed resistor can be substituted in place of the potentiometer (R2) in this circuit. This resistance affects the average amplitude of the generated noise signal, and, to some extent, the "color." Strictly speaking, however, this circuit generates white noise. If your intended application calls for the deeper-sounding pink noise, you can pass the output of this circuit through a low-pass filter circuit with a very gradual cut-off slope.

A better white noise generator circuit is shown in Fig. 5-8. This one uses the base-emitter junction of an npn transistor as the noise source. Notice that no connection is made to the transistor's collector at all. Almost any low-power npn transistor should work with this circuit, although you might need to experiment with several units until you find the transistor with the noisiest base-emitter junction. That's the one that will work the best for this application, even though it might be rejected for most other applications.

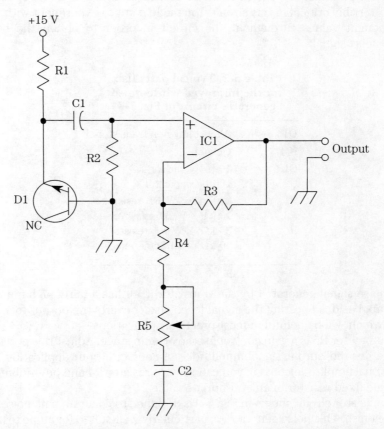

5-8 This improved white-noise generator circuit uses the base-emitter junction of an npn transistor as the noise source.

In this sort of application, the noise transistor's base-emitter junction is reverse-biased and is forced into zener breakdown. The voltage at which this happens will depend on the specific transistor type used, and it will typically be somewhere in the neighborhood of 7 to 8 volts. This zener breakdown effect is what actually generates the noise signal.

The op-amp IC serves to amplify the noise signal. The requirements for this op amp are as loose as they ever can be. After all, in an application like this, there is obviously no advantage at all in using an expensive low-noise op-amp chip—we want the noise. A common 741 op-amp IC would be more than adequate in this circuit. Notice that the op amp's power supply connections are not explicitly shown here, to avoid cluttering the schematic diagram. This is fairly standard practice. The appropriate power supply connections must always be assumed for any op-amp device. No op-amp circuit will work without the correct supply voltages, and the chip might be damaged if it is fed an input signal with no power supply.

A typical parts list for this white-noise generator circuit is given in Table 5-2. Nothing is terribly critical in this circuit. You might want to experiment with some of the component values throughout the circuit in order to obtain the best and strongest noise signal.

**Table 5-2. Typical parts list
for the improved white-noise
generator circuit of Fig. 5-8**

Q1	Almost any npn transistor
IC1	Op amp
C1	10-pF capacitor
C2	100-pF capacitor
R1	220-kΩ, ¼-W, 5% resistor
R2, R3	1 MΩ, ¼-W, 5% resistor
R4	33-kΩ, ¼-W, 5% resistor
R5	50-kΩ potentiometer

The noise signal generated by the transistor itself has a fairly high impedance. The op amp, besides boosting the signal level, also converts the output to a low impedance, which will be helpful in most practical applications.

Potentiometer R5 is a gain control for the op amp stage. Adjust this potentiometer for the desired output signal amplitude. Of course, if your application doesn't need a manual amplitude control, you can replace resistor R4 and potentiometer R5 with a single fixed resistor of an appropriate value.

The transistor circuit shown in Fig. 5-9 is designed to generate pink noise, rather than white noise. This noise generator is particularly well-suited for audio testing applications. A suitable parts list for this circuit is given in Table 5-3.

This pink-noise generator circuit uses two npn transistors. Q1 is the noise source. As in the previous circuit, this transistor is diode-connected, using just the base-emitter junction, leaving the collector floating. The second transistor is a simple audio amplifier. If it adds some extra noise of its own, so much the better. It is not absolutely essential to use the same type number for both transistors in this circuit, but it is a good idea. If the circuit does not generate enough of a noise signal, or produces noise only intermittently, try exchanging the two transistors. The odds are that one of them will have a sufficiently noisy base-emitter junction.

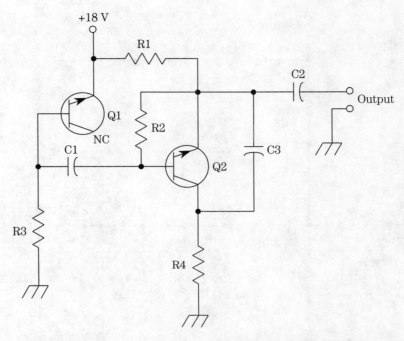

5-9 This transistor circuit is designed to generate pink noise, rather than white noise.

Table 5-3. Typical parts list for the pink-noise generator circuit of Fig. 5-9

Q1, Q2	npn transistor (2N2712, or similar)
C1, C2	0.1-μF capacitor
C3	0.0047-μF capacitor
R1	100-kΩ, ¼-W, 5% resistor
R2, R3	1 MΩ, ¼-W, 5% resistor
R4	100-Ω, ¼-W, 5% resistor

Capacitor C3, connected across the output, acts like a very crude low-pass filter. It de-emphasizes the higher frequency content of the noise signal, resulting in a fairly good approximation of pink noise. For different effects, you might want to try experimenting with different values for this capacitor.

The output impedance of this circuit is fairly high. An impedance-matching transformer or buffer amplifier stage might be required in some practical applications.

Project #7—Two-way noise generator

Figure 5-10 shows a deluxe noise-generator circuit. This one could be considered a sort of "noise function generator" (see chapter 4), because it has two signal

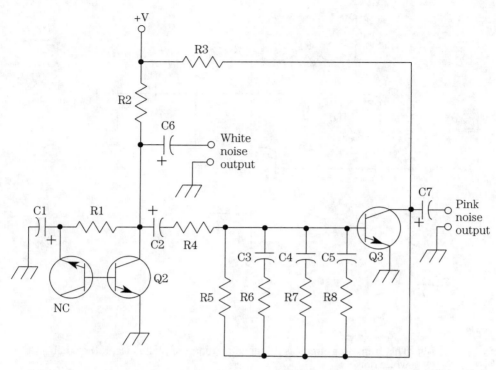

5-10   Project #7—Two-way noise generator.

outputs. It generates both white noise and pink noise. The two output signals can be used one at a time or simultaneously.

The supply voltage for this circuit can be anything from +15 volts to +30 volts. Naturally, the actual supply voltage used will have an affect on the amplitude of the output signals.

A typical parts list for this two-way noise-generator circuit is given in Table 5-4. Again, there is plenty of flexibility in most of the component values throughout this circuit. Feel free to experiment. This will only alter some of the characteristics of the noise signals generated.

Table 5-4. Suggested parts list for
Project #7—Two-way noise-generator circuit of Fig. 5-10

Q1, Q2, Q3	npn transistor (Radio Shack RS1617, GE-20,SK3020, or similar)
C1, C2	22-µF, 35-V electrolytic capacitor
C3	0.0047-µF capacitor
C4	0.0033-µF capacitor
C5	820-pF capacitor
C6, C7	1-µF, 35-V electrolytic capacitor
R1	62-kΩ, ¼-W, 5% resistor
R2, R3	6.2-kΩ, ¼-W, 5% resistor
R4	39-kΩ, ¼-W, 5% resistor

R5	1 MΩ, ¼-W, 5% resistor
R6	390-kΩ, ¼-W, 5% resistor
R7	100-kΩ, ¼-W, 5% resistor
R8	18-kΩ, ¼-W, 5% resistor

Once more, the noise source is the base-emitter junction of Q1, which has its collector left floating. The noise signal is amplified by transistor Q2 and fed to the white-noise output. The remainder of the circuit, from capacitor C2 on, is a moderately sophisticated low-pass filter network used to convert the original white-noise source into a pink-noise signal. Using the component values suggested in the parts list, the signal amplitude is dropped about 3 dB-per-octave.

Because some of the overall signal amplitude is subtracted by the filter network, transistor Q3 is added as a booster amplifier stage. This causes the overall signal level at the pink-noise output to be comparable to the signal level at the white-noise output.

The MM5837N/S2688 noise-generator IC

As with most other common electronics applications, dedicated ICs designed for use in noise generator circuits are available. One such device is the MM5837N, illustrated in Fig. 5-11. This is actually a digital device. The noise signal it generates is not truly random, but pseudo-random, though you'll never notice the difference in any practical application I can think of. Basically, the MM5837N generates a string of pseudo-random digital values (numbers) by a 17-bit shift register that is clocked by an internal digital oscillator.

Even though the MM5837N is packaged in a standard 8-pin DIP housing, it actually has only four active pins. Pins #5 through #8 are not internally connected to anything. They are simply ignored in practical circuits using this chip.

Three of the four active pins are for power supply connections. Pin #1 is used for the positive supply voltage, and should have a nominal value of +14 volts. There is a 2-volt leeway in either direction for this voltage, so the supply voltage can range anywhere from +12 volts to +16 volts. A standard +15-volt power supply would probably be a good choice for most practical applications. A +12-volt power supply should

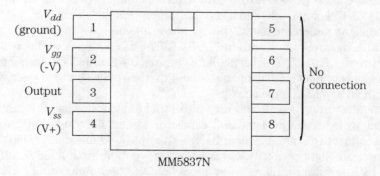

5-11   The MM5837N is a typical pseudo-random noise-generator IC.

work, but it might be a little less reliable if its voltage should ever happen to drop below its nominal value for any reason.

Pin #1 is labelled V_{ss} on the manufacturer's specification sheet, but this is just a fancy name for *ground connection*. Of course, the nominal voltage applied to this pin should be 0 volts.

The use of pin #2 (V_{gg}) is optional. If it is not used, the output signal will always be positive. Adding a negative supply voltage to the MM5837N permits its output signal to swing above and below the zero baseline. In most applications, this won't make much, if any, practical difference—but in some special cases it could matter, so it is nice to have the option. If this pin is used, the nominal V_{ss} voltage should be –27 volts. Again, there is a 2-volt leeway in either direction for this voltage, so the actual negative supply voltage can range anywhere from –25 volts to –29 volts. The fourth active pin on the MM5837N (pin #3) is simply the signal output pin. This is where the noise signal is tapped off.

Another dedicated noise-generator IC you might encounter is the S2688. This device is identical to the MM5837N. The difference in number is due to the fact that it was originated by a different manufacturer. There might be some minor internal differences in the on-chip circuitry of the MM5837N and the S2688, but they are functionally identical in all ways. Throughout this section, everything said about the MM5837N also applies to the S2688.

Being a digital device, the noise signal generated by the MM5837N is in the form of a series of numerical values. The sequence of these numbers is functionally random, but it does repeat, so it is only pseudo-random. A pseudo-random number generator produces a long jumbled pattern that it repeats endlessly. For example:

50842417395084241739508424173950842417395084241739 . . .

If the pattern is long enough, and the frequencies within the pattern are sufficiently randomized, it will be difficult, if not impossible to detect the repetition of the pattern. Certainly, it will not be possible to detect it by ear. Any practical pseudo-random generator, including the MM5837N, generates a much longer numerical sequence than in our example.

With only three or four active pins (remember, the use of pin #2 is optional), the MM5837N noise-generator IC is extremely easy to use. For example, Fig. 5-12 shows the basic MM5837N white-noise generator circuit. It is about as simple as any electronic circuit can be expected to be. No external components at all are required—just the MM5837N chip itself, and a suitable supply voltage source.

To adapt the generated signal to pink noise, all you have to do is add a passive low-pass filter network, as shown in Fig. 5-13. The exact values of this resistor and capacitor are not critical. The particular values used will affect the frequency spectrum of the output noise signal, of course, and therefore the sound of the noise when fed through an amplifier and speaker.

Try breadboarding this circuit, with the output fed to a small audio amplifier, and experiment with various resistor and capacitor values. Of course, as always, you should disconnect the circuit power before attempting to make any changes in the circuitry, or you might seriously damage the IC. I've found that good results can be obtained with resistors ranging from about 1 kΩ (1,000 ohms) to 12 kΩ (12,000 ohms) and capacitors ranging from 0.01 μF up to about 0.22 μF or so. You might

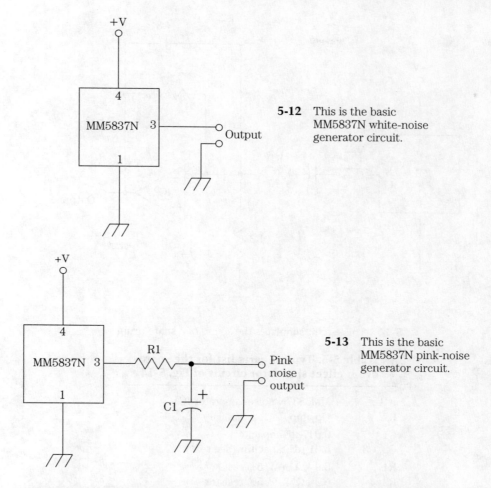

5-12 This is the basic MM5837N white-noise generator circuit.

5-13 This is the basic MM5837N pink-noise generator circuit.

want to replace the simple fixed resistor shown here with a 10-kΩ potentiometer in series with a 1-kΩ resistor.

A more sophisticated sound effect circuit using the MM5837N noise-generator IC is illustrated in Fig. 5-14. When normally open pushbutton switch S1 is briefly closed, a sound resembling a beat of a snare drum will be heard. A typical parts list for this circuit is given in Table 5-5. Experiment with other component values throughout this circuit to achieve different sound effects. Nothing is too critical here, but the values of capacitors C2 and C3 should always be identical. The circuit might not work reliably, if at all, if these capacitor values are mismatched. Assuming all other component values in the circuit are held constant, decreasing the value of capacitors C2 and C3 will result in a higher effective output frequency—it will sound like a smaller, higher-pitched drum. Not all combinations of component values in this circuit will sound much like a real drum, but the effect will always be percussive, and more or less drum-like.

Any standard op-amp IC can be used in this circuit. There is no point in using an

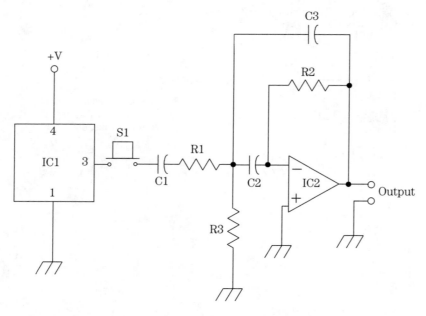

5-14   This circuit simulates the sound of a snare drum.

Table 5-5. Typical parts list for the snare drum effect simulator circuit of Fig. 5-14

IC1	MM5837N noise-generator IC
IC2	Op amp (741, or similar)
C1	0.047-μF capacitor
C2, C3	0.01-μF capacitor (see text)
R1	3.9-kΩ, ¼-W, 5% resistor
R2	6.2-kΩ, ¼-W, 5% resistor
R3	1-kΩ, ¼-W, 5% resistor
S1	SPST normally open pushbutton switch

expensive low-noise op amp chip in this sort of application. The power supply connections to the op amp itself are not shown in the schematic diagram, but they are always assumed. This is fairly standard practice.

You can build several copies of this circuit, each with its own output frequency, to create a full set of electronic drums. It is not necessary to use a separate MM5837N for each copy. One noise generator can drive several drum-effect circuits in parallel, as illustrated in Fig. 5-15.

Random voltage generation

A less common, but still interesting application for a pink-noise or white-noise generator is random (or pseudo-random) voltage generation. This type of applica-

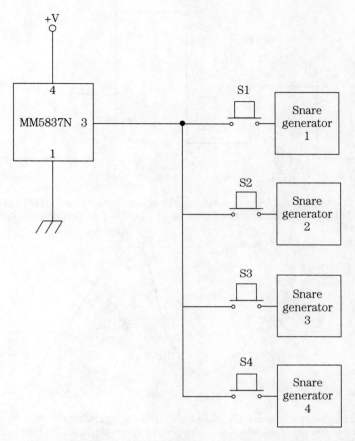

5-15   One noise generator can drive several drum effect circuits
in parallel.

tion is best suited for use with a digital noise source, such as the MM5837N noise-generator IC discussed above.

Basically, this device puts out a continuous stream of more or less randomized digital values. Changing digital signals can be transformed into a fluctuating analog signal by using a circuit known as a digital-to-analog converter, or D/A converter. The output of this circuit at any instant is an analog voltage that is proportional to the digital input signal at that particular moment, as illustrated in Fig. 5-16. A filter capacitor across the output is usually added to smoothly round off the edges of the sharp steps between values, as shown in Fig. 5-17.

In the MM5837N, the digital values being converted to analog voltages change at an extremely high rate, so all the individual values are blended together to create the noise signal—which is usually what we want. But in this particular application, we want to slow things down considerably, so that each individual voltage value is held for a noticeable period of time before it changes. There are several ways to do this.

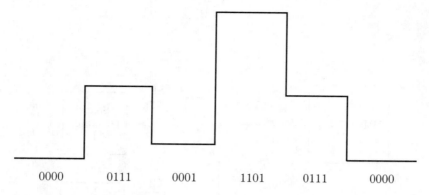

0000   0111   0001   1101   0111   0000

5-16 The output of a D/A converter circuit at any instant is an analog voltage that is proportional to the digital-input signal at that particular moment.

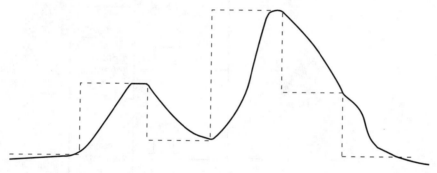

5-17 A filter capacitor is usually placed across the output of a D/A converter circuit to smoothly round off the edges of the sharp steps between values.

In a digital circuit, one way to slow down the changing values is to pass the digital noise signal through a counter of some sort. Instead of changing its output with every pulse, it changes its output every "X" pulses.

One application of this idea is illustrated in the simple yes/no circuit shown in Fig. 5-18. The output signal from the MM5837N noise generator is fed into a CD4027 JK flip-flop. It is used here as a simple divide-by-two counter. Normally, one of the two output LEDS (D1 or D2) will be lit, and the other will be dark. The flip-flop outputs these LEDs are connected to are always at opposite states—when one is HIGH, the other is LOW, and vice versa.

Only a single current-limiting resistor (R1) is needed for both LEDs, because they will never be simultaneously lit. Nothing much will happen in this circuit, because the input to the CD4027 is held LOW through resistor R2. When normally open pushbutton switch S1 is held closed, the output signal from the MM5837N noise generator can get through to the CD4027. Because of the large value of resistor R2, it doesn't do much while the switch is permitting the active signal to get through. The output from the MM5837N is a string of rapidly changing pulses. On each pulse, the flip-flop will reverse its output state. The LED that was previously dark is now lit, and vice versa.

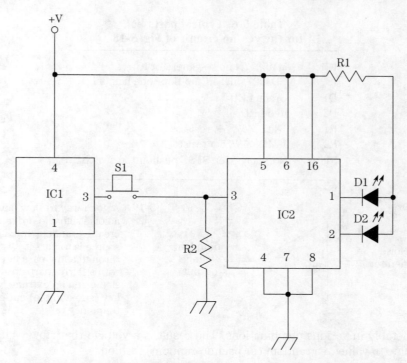

5-18 This simple yes/no circuit is based on a white-noise signal.

These changes occur at such a rapid rate that the human eye cannot distinguish between the individual blinks. Both LEDs will appear to be continuously lit as long as the switch is held closed. Actually, they are alternating on and off very rapidly.

When the pushbutton switch (S1) is released, it reverts to its normal open state. The signal path between the MM5837N and the CD4027 is broken, and the flip-flop's output goes LOW (grounded through resistor R2). The flip-flop's output will lock onto their present state. One of the LEDs will be lit and the other will be dark, but there is no way to predict which LED will be in LOW or HIGH state until it happens.

It is usually most effective to use two different colors of LEDs for maximum visible contrast. A good choice would be a green LED for D1 and a red LED for D2. If we label the green LED "HEADS," and the red LED "TAILS," the circuit can be used as an electronic "coin-flipper," or a novelty "executive decision maker."

A typical parts list for this circuit is given in Table 5-6. The exact resistor values are not critical. Changing the value of R1 alters the effective brightness of the LEDs when lit. Otherwise, changing the resistance values won't make very much difference—unless they are changed so much the circuit ceases to work at all.

Another way to slow down a noise signal to create a random voltage source is to pass it through a low-pass filter with a fairly sharp cut-off slope, and an extremely low (sub-audible) cut-off frequency—typically 5 Hz or less. This is illustrated in block diagram form in Fig. 5-19.

The high-frequency components are all deleted from the output signal, leaving only the slow-changing very low frequency components that occur irregularly and

**Table 5-6. Typical parts list
for the yes/no circuit of Fig. 5-18**

IC1	MM5837N noise-generator IC
IC2	CD4027 dual JK flip-flop (one half only)
D1	Green LED
D2	Red LED
R1	1-kΩ, ¼-W, 5% resistor
R2	1-MΩ, ¼-W, 5% resistor
S1	Normally open SPST pushbutton switch

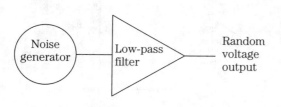

5-19 A good way to slow down a noise signal in order to create a random-voltage source is to pass it through a low-pass filter with a fairly sharp cut-off slope, and an extremely low (sub-audible) cut-off frequency.

unpredictably in various combinations. The result is a voltage that slowly changes from value to value in a random (or pseudo-random) fashion.

Project #8—Random voltage source

A practical random voltage source project using the filter method is shown in Fig. 5-20. A suitable parts list for this project appears as Table 5-7. The noise source in this project is just the MM5837N noise generator chip wired in its basic white noise generator configuration. Nothing at all complicated here.

The filter circuit is made up of two identical sections in series. Each is a second-order filter, so together they make up a fourth-order filter, which has a pretty sharp cut-off slope.

Because both filter sections are identical, we can look at just one at a time. One of the filter stages is shown by itself in Fig. 5-21. For this filter circuit to function properly, the following component equalities must be followed:

$$C_1 = C_2$$
$$R_1 = R_2$$
$$R_4 = 0.586R_3$$

The value of resistor R3 can be somewhat randomly selected, but the op amp's gain must be 1.586. The reasons for this are complicated and technical, so for now, you can just accept this fact as a given.

By using a 1.8-kΩ (1,800 ohms) resistor for R3, the nominal value for R4 should be:

$$R_4 = 0.586 \times 1800$$
$$= 1055 \text{ ohms}$$

We can round this off to 1 kΩ (1,000 ohms)—a standard, readily-available resistor value. It certainly should be close enough, because normal resistance tolerances are

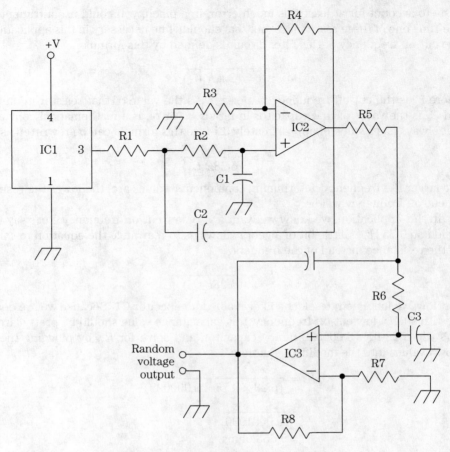

5-20   Project #8—Random voltage source.

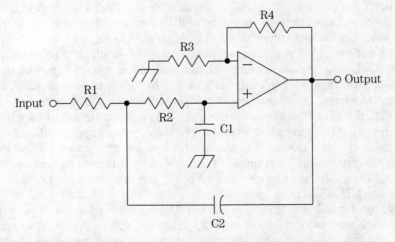

5-21   Both filter sections in the circuit of Fig. 5-20 are identical to this one.

likely to account for at least this much error. In a pinch, you could use a trimpot to fine-tune one of these resistances, but that shouldn't be necessary in this application. The cut-off frequency of this filter circuit is defined by this formula:

$$F_c = \frac{1}{(2\,\pi\,R_1 C_1)}$$

where F_c is the cut-off frequency in Hertz, R_1 is the value of that resistor in ohms, and C_1 is the value of that capacitor in farads. π, or pi, is a mathematical constant that always has a value of approximately 3.14, so this formula can be rewritten as:

$$F_c = \frac{1}{(6.28 R_1 C_1)}$$

The larger the frequency determining components' values are, the lower the resulting cut-off frequency will be.

In this application, we know we want a very low cut-off frequency. Let's say we decide on 0.75 Hz. A little bit of algebra allows us to rearrange the equation to solve for the resistance instead of the frequency:

$$R_1 = \frac{1}{(6.28 F_c C_1)}$$

Now, we just need to select a likely value for capacitor C1. Because we are dealing with such a low cut-off frequency, this capacitance value should be pretty large. Let's use a 22-μF (0.000022 farad) capacitor, and solve for R_1, by plugging these known values into the modified equation:

$$R_1 = \frac{1}{(6.28 \times 0.75 \times 0.000022)}$$

$$= \frac{1}{0.0001036}$$

$$= 9{,}653 \text{ ohms}$$

A commonly available 10-kΩ (10,000 ohms) resistor is close enough. After all, the exact cut-off frequency isn't all that important in this application. If for some reason you don't happen to have a 10-kΩ resistor handy, a 8.2-kΩ or 12-kΩ, or even 15-kΩ resistor would work pretty much as well. The second filter stage is identical to the first, so all component values are simply duplicated here.

Usually in filter circuits, it is a good idea to use high-grade, low-noise op amps— but in this application we are dealing with noise signals anyway, so such high-quality (and expensive) chips wouldn't offer any noticeable advantage. Common 741s will work every bit as well. A 747 dual op amp IC would also be a good choice.

As usual, the power supply connections to the op amps are not explicitly shown to avoid cluttering the schematic diagram. The appropriate supply voltages must be assumed in all op amp circuits. No op amp circuit will work without a suitable set of power supply connections, and the ICs are liable to be damaged or destroyed if you connect it incorrectly. Most modern op amp ICs can use the same positive supply voltage as the MM5837N noise-generator IC.

Finally, an extra-large filter capacitor (C5) is connected across the output to smooth out the voltage changes. In some applications, you might prefer to omit this capacitor. If used, its exact value isn't at all critical, or even important. Just so it's big—that's all that matters in this project.

Complex waveform generators

This chapter might well be the most fascinating and fun one in the book. So far we have been concentrating only on fairly basic, and quite simple waveforms. In some applications, however, we might need to generate more complex waveforms. A couple of obvious examples are imitative sound synthesis (electronically simulating the complex sound of a real-world musical instrument) and test pattern generation for television receivers. There are many other specialized applications for complex waveforms as well.

In this chapter, we will consider some ways of generating signals that are more complex than the usual sine waves, square waves, and so forth that we dealt with in previous chapter. Theoretically, the techniques described in this chapter should make it possible to electronically generate any desired waveform. In actual practice, however, there are real limits. Beyond a certain level of complexity, it becomes increasingly impractical to generate signals. The circuit becomes increasingly bulky, expensive, and difficult to build and use. Still, as long as you keep your expectations realistic, there is quite a lot you can do simply and inexpensively when it comes to generating unusual waveforms for all sorts of specialized purposes.

Amplitude modulation

Amplitude modulation, or *AM*, is a technique for combining two signals to create a new, complex composite signal. The primary signal is called the *carrier*, and the second signal is called the *program*. The program signal controls, or modulates, the instantaneous amplitude of the carrier signal. In other words, at any given moment, the instantaneous value of the program signal sets the positive- and negative-peak limits of the carrier waveform.

In most practical applications, the frequency of the program signal will be considerably less than the frequency of the carrier signal. Each complete cycle of the program signal controls several entire cycles of the carrier signal, as illustrated in Fig. 6-1.

171

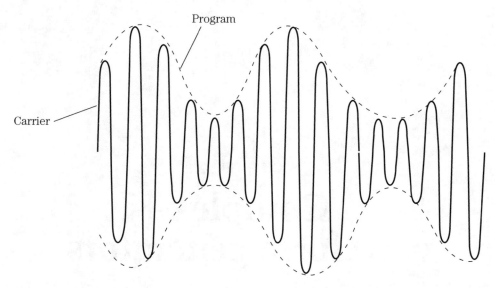

6-1 In amplitude modulation, each complete cycle of the program signal controls several entire cycles of the carrier signal.

At this point in our discussion, we are going to make a couple limiting assumptions. They won't always be true in practice, but they will simplify things for the present discussion. First, we will assume that all carrier and program signals are pure sine waves, with no harmonic content. Second, we will assume that the carrier signal is always an audio frequency signal, and that the modulated output signal is fed through a loudspeaker, so we can hear the results. Finally, we will assume that the program frequency is always less than the carrier frequency.

For now, we will set the carrier frequency at 1 kHz (1000 Hz). Let's first assume that the program frequency is very low, well below the audible range—say about 2 to 8 Hz. We will be able to hear the audible fluctuations in the output signal. This effect is known in music as *tremolo*.

As the program frequency is increased up towards the audible frequency range, we can no longer hear the individual fluctuations of the amplitude. Instead, we will hear a rather harsh, single tone, made up of the two original audio frequency signals, plus some newly-generated frequency components called *sidebands*. What we are now hearing is the effect of amplitude modulation.

Let's say we've increased the program frequency all the way up to 200 Hz. The carrier frequency is still 1000 Hz. The output signal will contain both of these frequency components. It will also include sidebands equal to the sum and difference of each of the original frequency components. In other words, the output signal in this example will be made up of the following four frequencies:

200 Hz	Program
800 Hz	Carrier – program
1000 Hz	Carrier
1200 Hz	Carrier + program

Changing either the program frequency or the carrier frequency (or both) will change the sideband frequencies. For example, let's leave the carrier signal at 1000 Hz, but increase the program frequency to 600 Hz. Now the output signal is made up of the following four frequency components:

400 Hz	Carrier – program
600 Hz	Program
1000 Hz	Carrier
1600 Hz	Carrier + program

Notice that the carrier – program sideband now has a lower frequency than the program signal itself. This will often happen. It doesn't change anything. Notice also that in most cases, these frequencies will not bear any particular harmonic relationship to each other. That is why the tone sounds so harsh, and can have a rather indefinite sense of pitch.

Using sine waves as the input signals will always result in two sidebands being produced during amplitude modulation. The amplitude of the two inputs signals doesn't affect the sideband frequencies, although it can affect the strength or amplitude of these newly-generated frequency components.

If either the program signal or the carrier signal (or both) is a more complex waveform, more sidebands will be generated. There will always be a pair (sum and difference) of generated sidebands for every frequency component in the carrier signal combined with every frequency component in the program signal.

Let's repeat the last example with a 600-Hz program signal amplitude modulating a 1000-Hz carrier signal—but this time the carrier signal is a 1000-Hz square wave, not a sine wave. The program signal is still a sine wave.

A square wave, you should recall, is comprised of the fundamental frequency, and all the odd harmonics, but no even harmonics. The relative amplitude of the harmonics are fairly strong. However, to avoid excessive clutter and confusion, we will ignore everything over the tenth harmonic. Therefore our carrier signal in this modified example is made up of the following frequency components:

1000 Hz	Fundamental
3000 Hz	Third harmonic
5000 Hz	Fifth harmonic
7000 Hz	Seventh harmonic
9000 Hz	Ninth harmonic

The program signal is still a sine wave, so it consists of only its fundamental frequency, 600 Hz.

The program signal will generate a sideband pair with each frequency component in the carrier signal, so the resulting amplitude modulated output signal will be made up of the following frequency components:

400 Hz	Carrier (fundamental) – program
600 Hz	Program
1000 Hz	Carrier (fundamental)
1600 Hz	Carrier (fundamental) + program
2400 Hz	Carrier (third) – program

3000 Hz   Carrier (third)
3600 Hz   Carrier (third) + program
4400 Hz   Carrier (fifth) – program
5000 Hz   Carrier (fifth)
5600 Hz   Carrier (fifth) + program
6400 Hz   Carrier (seventh) – program
7000 Hz   Carrier (seventh)
7600 Hz   Carrier (seventh) + program
8400 Hz   Carrier (ninth) – program
9000 Hz   Carrier (ninth)
9600 Hz   Carrier (ninth) + program

If the program signal is also a multi-frequency component waveform (anything more complex than a sine wave), every frequency component in the program signal will produce a pair of sidebands with every frequency component in the carrier signal. As you can well imagine, there are soon a LOT of sidebands, and the signal gets very, very complex.

To avoid making the example too long, we will ignore everything above the fifth harmonic this time. The carrier signal is a 1000-Hz square wave with the following frequency components:

1000 Hz   Fundamental
3000 Hz   Third harmonic
5000 Hz   Fifth harmonic

The program signal in this example is a 600-Hz square wave, comprised of the following frequency components:

600 Hz   Fundamental
1800 Hz   Third harmonic
3000 Hz   Fifth harmonic

The resulting amplitude-modulated output signal of these two inputs is summarized in Table 6-1. Notice how complex things have gotten, and we're not even considering anything above the fifth harmonics. Also notice that in some cases, the program harmonic frequency is higher than the carrier frequency component. The difference is still assumed to be a positive value. The smaller frequency is always subtracted from the larger frequency, to give the absolute value of the difference. Notice also that the difference of the carrier third harmonic and the program fifth harmonic is 0. This happens occasionally. That sideband is simply cancelled out in the output signal of this particular combination.

**Table 6-1. The frequency
components of the amplitude
modulation example in the text**

0 Hz	$(C_3 - P_5)$
400 Hz	$(C_F - P_F)$
600 Hz	(P_F)

800 Hz	$P_3 - C_F$)
1000 Hz	(C_F)
1200 Hz	($C_3 - P_3$)
1600 Hz	($C_F + P_F$)
1800 Hz	(P_3)
2000 Hz	($P_5 - C_F$) also ($C_5 - P_5$)
2400 Hz	($C_3 - P_F$)
2800 Hz	($C_F + P_3$)
3000 Hz	(C_3) also (P_5)
3200 Hz	($C_5 - P_3$)
3600 Hz	($C_3 + P_F$)
4000 Hz	($C_F + P_5$)
4400 Hz	($C_5 - P_F$)
4800 Hz	($C_3 + P_3$)
5000 Hz	(C_5)
5600 Hz	($C_5 + P_F$)
6000 Hz	($C_3 + P_5$)
6800 Hz	($C_5 + P_3$)
8000 Hz	($C_5 + P_5$)

C_F = Carrier fundamental
C_3 = Carrier third harmonic
C_5 = Carrier fifth harmonic
P_F = Program fundamental
P_3 = Program third harmonic
P_5 = Program fifth harmonic

To produce amplitude modulation, you need two oscillators (or other signal sources) and a special circuit known as a VCA, or voltage-controlled amplifier. The gain of a VCA is determined by the value of a control voltage. The carrier signal is used as the VCA's signal input, and the program signal serves as the control voltage, as illustrated in Fig. 6-2.

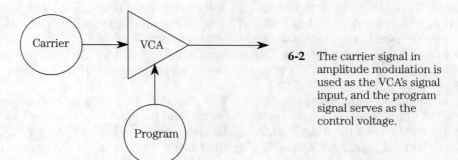

6-2 The carrier signal in amplitude modulation is used as the VCA's signal input, and the program signal serves as the control voltage.

Ring modulation

Ring modulation is similar to amplitude modulation, except the original program and carrier signals are suppressed from the output. The output signal is made up solely of the generated sidebands.

For example, let's assume we have a 435-Hz program signal modulating a 980-Hz carrier signal, and both signals are pure sine waves. If we use amplitude modulation, the resulting output signal will be made up of the following four frequency components:

435 Hz	Program
545 Hz	Carrier – program
980 Hz	Carrier
1415 Hz	Carrier + program

But if we subject these same input signals to ring modulation, the output signal will be made up of just these two frequency components:

545 Hz	Carrier – program
1415 Hz	Carrier + program

Not surprisingly, none of the frequency components in a ring-modulated signal are likely to be at all harmonically related to each other, although a few stray harmonics could show up with certain combinations of carrier and program signals.

Of course, if either the carrier signal or the program signal (or both) is more complex than a sine wave, there will be more sidebands in the ring-modulated output signal, just as was the case with amplitude modulation.

Frequency modulation

Another common form of signal modulation that can be used to create complex waveforms is *frequency modulation*, or *FM*. Like amplitude modulation (AM), frequency modulation combines two input signals called the carrier and the program and generate new frequency components called sidebands. However, a FM system produces sidebands differently than an AM system.

You have probably noticed that the terms AM and FM are the same ones used for types of radio broadcasts. Yes, it is the same thing. In radio broadcasts, the program signal is the audio material to be broadcast—the talking and music. The carrier signal is a radio frequency (RF) sine wave. This is the frequency the radio is tuned to in order to receive the desired station and program material.

In amplitude modulation, the carrier signal is fed through a VCA (voltage-controlled amplifier), whose gain is varied by the program signal, and is used as a control voltage.

In frequency modulation, there is no VCA, but the carrier signal source must be a VCO (voltage-controlled oscillator). Again, the program signal is used as a control voltage, but this time it controls the signal frequency of the carrier VCO, rather than the signal amplitude. A simplified block diagram of an FM system is shown in Fig. 6-3.

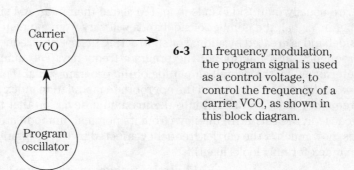

6-3 In frequency modulation, the program signal is used as a control voltage, to control the frequency of a carrier VCO, as shown in this block diagram.

Obviously, the carrier signal's frequency is no longer a constant. It is always varying from instant to instant, depending on the instantaneous amplitude of the program signal. A simplified frequency-modulated signal is illustrated in Fig. 6-4.

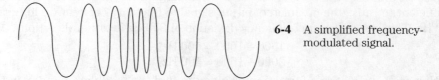

6-4 A simplified frequency-modulated signal.

Once again, we will start out by setting a few limiting assumptions to simplify our discussion. Unless otherwise noted, all the signals we will be working with are pure sine waves, with no harmonic content at all (just a single-frequency component—the fundamental). The carrier signal is an audio-frequency signal, and the frequency-modulated output signal is assumed to be fed out through an amplifier and loudspeaker, so it can be heard as a tone. Normally, the program signal's frequency is lower than the carrier signal's frequency.

If the program signal frequency is very low, say, 2 Hz or less, we will be able to actually hear its waveshape, as the output tone gradually changes from frequency to frequency.

If the program signal frequency is increased, but is still kept below the audible range—say 5 Hz to 10 Hz—the specific frequency changes will no longer be distinctly audible, but we will still be able to hear that the output tone is fluctuating in frequency. If the fluctuations are not too strong, this effect can be quite pleasing to the ear. In music, it is called *vibrato*. Electronically-generated tones often sound dull and lifeless with no vibrato. However, too much vibrato can sound terrible, and make musical notes sound badly out of tune. The depth of the vibrato effect is determined by the overall amplitude of the program signal waveform used to modulate the frequency of the carrier VCO.

When the program signal frequency is brought up to the audible range, (above 20 Hz, or so), things start to get interesting. True FM effects set in. Sidebands (new frequency components that are not part of either input signal) are generated, but they are not the same as the sidebands generated by amplitude modulation.

There are usually more sidebands in an FM signal than in an AM signal. With a varying program signal, the number of sidebands will vary from moment to moment.

Each sideband is created by adding or subtracting the carrier frequency and the program frequency, or a multiple of the program frequency. The number of sidebands is determined by the overall amplitude of the program signal. This number is called the *modulation index*. To find the appropriate modulation index number, the program frequency must be divided into the maximum deviation that the program signal will cause in the carrier frequency (from its nominal, unmodulated value). In other words, how much is the carrier frequency affected by the program signal? The modulation index formula looks like this:

$$Modulation\ index = \frac{(Maximum\ frequency\ deviation)}{(Program\ frequency)}$$

This sounds terribly complicated, but a few examples should make it very easy to understand. Let's assume that the carrier signal is a 1500-Hz sine wave, and we are going to frequency-modulate it with a 100-Hz sine wave as the program signal. We will also assume that the program signal is a constant, unvarying tone. Its amplitude is sufficient to cause the modulated output signal to fluctuate between 1000 Hz and 2000 Hz. This is a maximum frequency deviation of 500 Hz in either direction:

$$1500 - 1000 = 500\ Hz$$
$$2000 - 1500 = 500\ Hz$$

Now we have all the information we need to find the modulation index in this example:

$$Modulation\ index = \frac{(Maximum\ frequency\ deviation)}{(Program\ frequency)}$$
$$= \frac{500}{100}$$
$$= 5$$

This means that the frequency modulated output signal will contain five sidebands above the nominal carrier frequency, and five sidebands below it. The number of sidebands is always symmetrical around the nominal (unmodulated) carrier frequency. Each sideband will be removed from the nominal carrier frequency by a factor equal to the program frequency (100 Hz in our example). That is:

First upper sideband	$= F_c + (1 \times F_p)$
Second upper sideband	$= F_c + (2 \times F_p)$
Third upper sideband	$= F_c + (3 \times F_p)$
. . .	
Nth upper sideband	$= F_c + (N \times F_p)$

and:

First lower sideband	$= F_c - (1 \times F_p)$
Second lower sideband	$= F_c - (2 \times F_p)$
Third lower sideband	$= F_c - (3 \times F_p)$
. . .	
Nth lower sideband	$= F_c - (N \times F_p)$

where F_c is always the nominal (unmodulated) carrier frequency, and F_p is the program frequency.

The modulation index in our example is five, so the modulated output signal will have ten sidebands. The frequency components in the output signal will be:

Fifth lower sideband	1000 Hz	$(1500 - (5 \times 100))$
Fourth lower sideband	1100 Hz	$(1500 - (4 \times 100))$
Third lower sideband	1200 Hz	$(1500 - (3 \times 100))$
Second lower sideband	1300 Hz	$(1500 - (2 \times 100))$
First lower sideband	1400 Hz	$(1500 - (1 \times 100))$
Nominal carrier frequency	1500 Hz	
First upper sideband	1600 Hz	$(1500 + (1 \times 100))$
Second upper sideband	1700 Hz	$(1500 + (2 \times 100))$
Third upper sideband	1800 Hz	$(1500 + (3 \times 100))$
Fourth upper sideband	1900 Hz	$(1500 + (4 \times 100))$
Fifth upper sideband	2000 Hz	$(1500 + (5 \times 100))$

For our second example, let's leave the program signal exactly as it is, but increase the carrier frequency to 2250 Hz (2.25 kHz). The program signal has not been altered, so the modulation index and the maximum frequency deviation are the same as in the earlier example. The carrier signal itself does not affect these factors.

The maximum frequency deviation is still 500 Hz, so the carrier's actual (modulated) frequency will vary from a low of:

$$2250 - 500 = 1750 \text{ Hz}$$

to a high of:

$$2250 + 500 = 2750 \text{ Hz}$$

The frequency components of the modulated output signal in this example work out like this:

Fifth lower sideband	1750 Hz	$(2250 - (5 \times 100))$
Fourth lower sideband	1850 Hz	$(2250 - (4 \times 100))$
Third lower sideband	1950 Hz	$(2250 - (3 \times 100))$
Second lower sideband	2050 Hz	$(2250 - (2 \times 100))$
First lower sideband	2150 Hz	$(2250 - (1 \times 100))$
Nominal carrier frequency	2250 Hz	
First upper sideband	2350 Hz	$(2250 + (1 \times 100))$
Second upper sideband	2450 Hz	$(2250 + (2 \times 100))$
Third upper sideband	2550 Hz	$(2250 + (3 \times 100))$
Fourth upper sideband	2650 Hz	$(2250 + (4 \times 100))$
Fifth upper sideband	2750 Hz	$(2250 + (5 \times 100))$

If we keep the carrier signal the same (2250 Hz), but change the program frequency to 250 Hz—while keeping the amplitude of the program signal exactly the same—the maximum frequency deviation will still be 500 Hz (for a range of 2250 Hz to 2750 Hz), but the modulation index will be changed to:

$$Modulation\ index = \frac{(Maximum\ frequency\ deviation)}{(Program\ frequency)}$$

$$= \frac{500}{250}$$

$$= 2$$

Now there will be just two upper sidebands and two lower sidebands, giving us a modulated output signal with the following frequency components:

Second lower sideband	1750 Hz	$(2250 - (2 \times 250))$
First lower sideband	2000 Hz	$(2250 - (1 \times 250))$
Nominal carrier frequency	2250 Hz	
First upper sideband	2500 Hz	$(2250 + (1 \times 250))$
Second upper sideband	2750 Hz	$(2250 + (2 \times 250))$

For our last example, we will keep everything exactly the same as in the last example, except we will increase the amplitude of the program signal so that the maximum frequency deviation is increased to 1500 Hz. This means that the actual carrier frequency will be modulated from a low of:

$$2250 - 1500 = 750 \text{ Hz}$$

to a high of:

$$2250 + 1500 = 3750 \text{ Hz}$$

This increased maximum frequency deviation means the modulation index is changed to:

$$Modulation\ index = \frac{(Maximum\ frequency\ deviation)}{(Program\ frequency)}$$

$$= \frac{1500}{250}$$

$$= 6$$

The frequency-modulated output signal in this example will include the following frequency components:

Sixth lower sideband	750 Hz	$(2250 - (6 \times 250))$
Fifth lower sideband	1000 Hz	$(2250 - (5 \times 250))$
Fourth lower sideband	1250 Hz	$(2250 - (4 \times 250))$
Third lower sideband	1500 Hz	$(2250 - (3 \times 250))$
Second lower sideband	1750 Hz	$(2250 - (2 \times 250))$
First lower sideband	2000 Hz	$(2250 - (1 \times 250))$
Nominal carrier frequency	2250 Hz	
First upper sideband	2500 Hz	$(2250 + (1 \times 250))$
Second upper sideband	2750 Hz	$(2250 + (2 \times 250))$
Third upper sideband	3000 Hz	$(2250 + (3 \times 250))$
Fourth upper sideband	3250 Hz	$(2250 + (4 \times 250))$
Fifth upper sideband	3500 Hz	$(2250 + (5 \times 250))$
Sixth upper sideband	3750 Hz	$(2250 + (6 \times 250))$

Changing either the program signal's frequency or its amplitude changes the modulation index, and therefore, the number of sidebands in the final output signal.

Notice also that in each case, the result of subtracting the maximum frequency deviation from the nominal (unmodulated) carrier frequency is equal to the lowest sideband frequency, and the result of adding the maximum frequency deviation to the nominal (unmodulated) carrier frequency is equal to the highest sideband frequency. Frequency modulation always works out this way.

Another interesting effect of frequency modulation is that as the amount of modulation (the number of sidebands) is increased, the strength of the original carrier signal in the output is decreased. This, combined with the almost inevitably non-harmonic relationship can result in a strongly modulated signal losing all definite sense of pitch.

If either the carrier signal or the program signal (or both) is a more complex signal than a sine wave, the result of frequency modulation will be much more complex. Every frequency component (fundamental, harmonics, or whatever) in either or both of the input signals will produce its own complete set of sidebands in the modulated output signal. For most practical purposes, it is advisable to make at least one of the input signals to an FM system a clean sine wave, or the output signal will probably turn out too jumbled and muddy to be of much practical use at all.

Additive and subtractive synthesis

Amplitude modulation, ring modulation, and frequency modulation can produce some very unusual signals, but they are rarely true periodic waveforms, because of their many strong inharmonic frequency components. It is rare that a sideband will have a harmonic relationship to any other frequency component in the modulated signal. It does happen sometimes, but not very often. Because the sideband frequencies are not exact multiples, the shape of the wave will alter from cycle to cycle. This is due to the fact that the various inharmonic frequency components do not always begin and end their cycles at the same time. They don't have cycles of compatible lengths.

Besides modulation, complex waveforms can be formed by additive synthesis. That is, multiple relatively simple waveforms are combined into a single complex signal. Theoretically, any conceivable waveform or signal can be built up from sine waves of appropriate frequencies, using additive synthesis. In practice, however, this would be reasonable only if we were building up a relatively simple waveform, with not too many frequency components. If the desired signal is very complex, an impractical number of sine wave oscillators would be required—not to mention the difficulty of keeping them all tuned together, at the correct amplitudes, and with the proper phase relationships.

Additive synthesis does not have to be limited to using just sine waves as its building blocks. Any waveform or other signal you might have available can be used to additively synthesize a more complex signal.

Let's say we have three rectangle waves we are going to combine by additive synthesis. Signals X and Z are square waves, each with a duty cycle of 1:2. Signal Y is a rectangle wave with a duty cycle of 1:3. Signals X and Y both have the same amplitude, or signal strength, which we will call "A." The amplitude of signal Z is exactly half of either signal X or signal Y, or A/2.

To avoid getting overwhelmed by the multiple harmonics in all these rectangle waves, we will assume that any frequency component with an amplitude of less than A/15 is too weak to be of significance.

Signal X has a duty cycle of 1:2, and a fundamental frequency of 300 Hz, so its significant harmonic make-up looks like this:

300 Hz	Fundamental	A
900 Hz	Third harmonic	$\dfrac{A}{3}$
1500 Hz	Fifth harmonic	$\dfrac{A}{5}$
2100 Hz	Seventh harmonic	$\dfrac{A}{7}$
2700 Hz	Ninth harmonic	$\dfrac{A}{9}$
3300 Hz	Eleventh harmonic	$\dfrac{A}{11}$
3900 Hz	Thirteenth harmonic	$\dfrac{A}{13}$
4500 Hz	Fifteenth harmonic	$\dfrac{A}{15}$

Signal Y has a fundamental frequency of 600 Hz, and a duty cycle of 1:3, so its significant harmonic make-up looks like this:

600 Hz	Fundamental	A
1200 Hz	Second harmonic	$\dfrac{A}{2}$
2400 Hz	Fourth harmonic	$\dfrac{A}{4}$
3000 Hz	Fifth harmonic	$\dfrac{A}{5}$
4200 Hz	Seventh harmonic	$\dfrac{A}{7}$
4800 Hz	Eighth harmonic	$\dfrac{A}{8}$
6600 Hz	Eleventh harmonic	$\dfrac{A}{11}$
7800 Hz	Thirteenth harmonic	$\dfrac{A}{13}$
8400 Hz	Fourteenth harmonic	$\dfrac{A}{14}$

Finally, signal Z (with half the amplitude of the others) is a 150-Hz square wave. Notice that the fundamental frequency of each of these signals are harmonically related. The relevant frequency components of signal Z are:

150 Hz	Fundamental	$\dfrac{A}{2}$
450 Hz	Third harmonic	$\dfrac{A}{6}$
750 Hz	Fifth harmonic	$\dfrac{A}{10}$
1050 Hz	Seventh harmonic	$\dfrac{A}{14}$

When we combine these three rectangle waves with additive synthesis, we get a complex signal comprised of the following frequency components:

150 Hz	$\dfrac{A}{2}$
300 Hz	A
450 Hz	$\dfrac{A}{6}$
600 Hz	A
750 Hz	$\dfrac{A}{10}$
900 Hz	$\dfrac{A}{3}$
1050 Hz	$\dfrac{A}{14}$
1200 Hz	$\dfrac{A}{2}$
1500 Hz	$\dfrac{A}{5}$
2100 Hz	$\dfrac{A}{7}$
2400 Hz	$\dfrac{A}{4}$
2700 Hz	$\dfrac{A}{9}$
3000 Hz	$\dfrac{A}{5}$
3300 Hz	$\dfrac{A}{11}$
3900 Hz	$\dfrac{A}{13}$
4200 Hz	$\dfrac{A}{7}$
4500 Hz	$\dfrac{A}{15}$

$$4800 \text{ Hz} \quad \frac{A}{8}$$

$$6600 \text{ Hz} \quad \frac{A}{11}$$

$$7800 \text{ Hz} \quad \frac{A}{13}$$

$$8400 \text{ Hz} \quad \frac{A}{14}$$

Notice that because all three original fundamental frequencies were harmonically related, all frequency components in the combined signal are harmonics—if we assume 150 Hz is the new overall fundamental, even though it is not the highest amplitude frequency component, as a true fundamental usually is. Also notice the unusual and irregular structure of relative amplitudes as the frequency components increase in frequency. This is a typical result of additive synthesis.

To summarize, in additive synthesis, we start out with relative simple input signals and combine them into a more complex output signal. We simply add in the frequency components we want. Not surprisingly, subtractive synthesis works in just the opposite way. In this case, we start out with a relatively complex input signal and remove the frequency components we don't want, leaving us with a simpler output signal.

Subtractive synthesis is performed with frequency-sensitive circuits called *filters*. It is not appropriate to go into major detail on filters in this book. We will just touch upon the most basic principles here. If you are interested in more specific information on filter circuit, I suggest you read my earlier book, *Designing and Building Electronic Filters*.

There are four basic filter types. They are classified according to which frequency components they remove. The four basic types of filters are:

- Low-pass filter
- High-pass filter
- Band-pass filter
- Band-reject filter

The names of the four filter types are more or less self-explanatory. Other filter types are possible, and occasionally encountered, but they are extremely rare in most practical analog circuitry.

In all of the following examples, we will assume the input signal is a 300-Hz square wave, comprised of the following frequency components:

300 Hz	Fundamental
900 Hz	Third harmonic
1500 Hz	Fifth harmonic
2100 Hz	Seventh harmonic
2700 Hz	Ninth harmonic
3300 Hz	Eleventh harmonic
3900 Hz	Thirteenth harmonic
4500 Hz	Fifteenth harmonic

We will also be assuming that all of our filters are perfect, with infinitely sharp cut-off slopes.

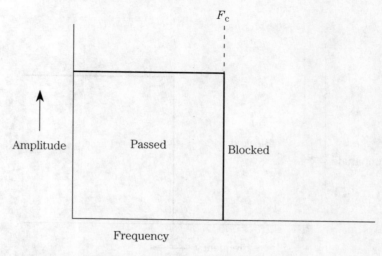

6-5   A typical frequency-response graph of an ideal low-pass filter.

A low-pass filter, as the name suggests, passes low frequency components, but blocks high frequency components. The switchover point is called the cut-off frequency. The frequency response graph of an ideal low-pass filter is shown in Fig. 6-5.

If the cut-off frequency is 3500 Hz, anything above the eleventh harmonic in our sample input signal (listed above) is deleted, leaving only the following frequency components:

300 Hz	Fundamental
900 Hz	Third harmonic
1500 Hz	Fifth harmonic
2100 Hz	Seventh harmonic
2700 Hz	Ninth harmonic
3300 Hz	Eleventh harmonic

Lowering the cut-off frequency to 2000 Hz, only the fifth harmonic and lower will be left:

300 Hz	Fundamental
900 Hz	Third harmonic
1500 Hz	Fifth harmonic

A high-pass filter is the exact opposite of a low-pass filter. In this case, everything below the cut-off frequency is deleted, and only those frequency components higher than the cut-off frequency remains in the output signal. A frequency-response graph for an ideal high-pass filter is illustrated in Fig. 6-6.

Feeding our example input signal into a high-pass filter with an ideal cut-off frequency of 3500 Hz leaves only these frequency components:

3900 Hz	Thirteenth harmonic
4500 Hz	Fifteenth harmonic

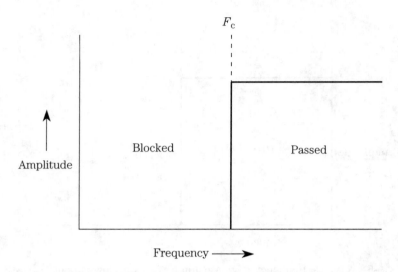

6-6   A typical frequency-response graph of an ideal high-pass filter.

With a cut-off frequency of 2000 Hz, the output signal will contain the following frequency components:

2100 Hz	Seventh harmonic
2700 Hz	Ninth harmonic
3300 Hz	Eleventh harmonic
3900 Hz	Thirteenth harmonic
4500 Hz	Fifteenth harmonic

Notice that using a high-pass filter breaks down the harmonic relationship of the frequency components, because the fundamental (the lowest frequency component) is the deleted if anything at all is removed.

A band-pass filter has two cut-off frequencies—a lower cut-off frequency and an upper cut-off frequency. The frequencies between these two comprise the passband. The output includes only those frequency components in the input signal that fall within this passband. Anything below the lower cut-off frequency, or above the upper cut-off frequency, is deleted from the output signal. A frequency-response graph for an ideal band-pass filter is illustrated in Fig. 6-7.

The distance between the lower cut-off frequency and the upper cut-off frequency is called the bandwidth. For example, let's imagine an ideal band-pass filter with a lower cut-off frequency of 1200 Hz and an upper cut-off frequency of 3000 Hz. The bandwidth of this filter is:

BW = Upper cut-off frequency – lower cut-off frequency
    = 3000 – 1200 Hz
    = 1800 Hz

Using our same sample input signal, defined above, the output signal from this ideal band-pass filter will include only the following frequency components:

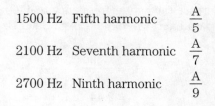

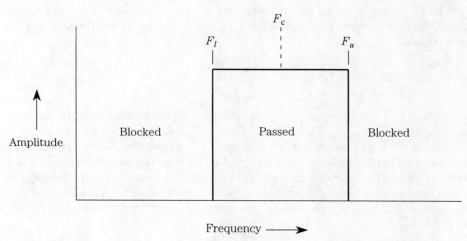

6-7 A typical frequency-response graph of an ideal band-pass filter.

In practical usage, a band-pass filter's specifications are usually given in terms of the center frequency (F_c) and Q (or quality factor).

The center frequency is the frequency at the exact midpoint between the lower cut-off frequency and the upper cut-off frequency. In our present example, the center frequency is 2100 Hz.

The Q, or "quality factor" has nothing to do with "good" or "bad." In some applications, a filter with a low Q will be much better than one with a high Q, while in other applications, the exact opposite might be true. It depends entirely on the specific requirements of the particular application at hand. The Q of a band-pass filter, is basically just an alternate, relative measurement of the bandwidth. The formula for finding the Q of a band-pass filter is:

$$Q = \frac{F_c}{BW}$$

F_c is the center frequency, and BW is the absolute bandwidth, both in Hertz. The units for Q values are undefined. It's just Q. In our example, the Q is:

$$Q = \frac{2100}{1800}$$

$$= 1.17$$

This is a fairly low value of Q.

Suppose that we increase the center frequency to 4400 Hz, but keep the bandwidth at 1800 Hz (the pass band will extend from 3500 Hz to 5300 Hz). This will change the Q to:

$$Q = \frac{4400}{1800}$$

$$= 2.44$$

Now, let's try a narrower bandwidth, say, only 500 Hz. The center frequency remains at 4400 Hz (the passband extends from 4150 Hz to 4650 Hz.) Now the Q of our filter works out to:

$$Q = \frac{4400}{500}$$

$$= 8.8$$

Increasing the center frequency, or decreasing the bandwidth (or both), increases the Q, or vice versa.

Just as the high-pass filter is the reverse of a low-pass filter, a band-reject filter is the functional opposite of a band-pass filter. Again, a specific band of frequencies is defined in exactly the same way as in a band-pass filter. A frequency response graph for an ideal band-reject filter is illustrated in Fig. 6-8. Because a notch is taken out of the frequency response, this type of filter is also frequently referred to as a *notch filter*.

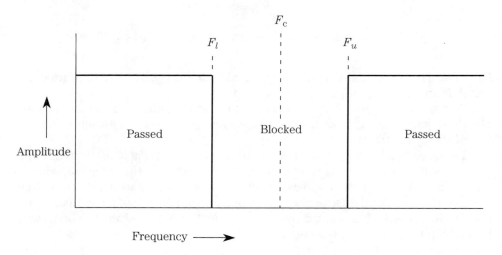

6-8   A typical frequency-response graph of an ideal band-reject filter.

Let's assume we have an ideal band-reject filter, with the following specifications:

Center frequency	=	3400 Hz
Q	=	5
Bandwidth	=	680 Hz
Lower cut-off frequency	=	2720 Hz
Upper cut-off frequency	=	4080 Hz

Using our same 300-Hz square-wave input signal, the output signal will include the following frequency components:

300 Hz	Fundamental
900 Hz	Third harmonic
1500 Hz	Fifth harmonic
2100 Hz	Seventh harmonic
2700 Hz	Ninth harmonic
4500 Hz	Fifteenth harmonic

The various filter types can be combined to create more complex subtractive synthesis.

Also, there is no reason why additive synthesis and subtractive synthesis can't be combined. For example, we could start out with the output signal from our three rectangle wave additive synthesis example described above, and then subject it to subtractive synthesis by passing the complex additive synthesis signal through a band-pass filter with the following specifications:

Center frequency	=	2600 Hz
Q	=	2.1
Bandwidth	=	1250 Hz
Lower cut-off frequency	=	1350 Hz
Upper cut-off frequency	=	3850 Hz

This leaves us with a combined additive/subtractive synthesis output signal made up of the following frequency components:

1500 Hz
2100 Hz
2400 Hz
2700 Hz
3000 Hz
3300 Hz

Throughout our discussion of subtractive synthesis, we have been assuming ideal filters, with infinitely sharp cut-off slopes. If the cut-off frequency of a low-pass filter is 1000 Hz, we have been assuming that any frequency component with a frequency of 999 Hz or less will be fully passed with no attenuation from the filter, while any frequency component with a frequency of 1001 Hz or higher is completely blocked (attenuated 100%). No practical filter circuit can achieve this ideal. A practical filter circuit has a rounded, more gradual cut-off slope, as indicated in the frequency response graph of Fig. 6-9. Frequency components close to the cut-off frequency are partially attenuated, somewhere in between completely passed and completely blocked.

Generally speaking, the steeper the cut-off (or roll-off) slope, the better the filter. A filter's cut-off slope is specified as the number of decibels per octave. An octave is a doubling of frequency. A low-pass filter with a 6-dB-per-octave roll-off will add another 6 dB of attenuation, each time the frequency is doubled.

As we have seen, additive synthesis and most types of subtractive synthesis also results in non-harmonic frequency components, so these are usually not true periodic waveforms. In the remainder of this chapter, we will consider ways of generating unusual periodic waveforms more or less directly.

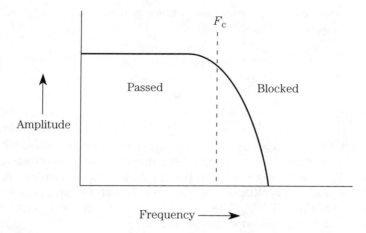

6-9 A typical frequency-response graph of a practical low-pass filter.

In some of the function generator circuits of chapter 4, a control was used to adjust the duty cycle of the rectangle wave output. These controls also affected the slew of the other waveforms (sine wave, triangle wave, sawtooth wave). Usually this is undesirable, except perhaps at extremes—at one end of the slew range, you might get a triangle wave, but at the opposite end of the control's range, you'd get a sawtooth wave. At intermediate settings, you would get a strange combination of a triangle wave and sawtooth wave, rather like the one illustrated in Fig. 6-10. This unusual, unnamed waveform might be just what you want in certain specialized applications. Sound synthesis is probably the most probable area for such things. A slewed sine wave can really be weird.

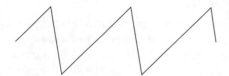

6-10 At intermediate settings of a "slew" control, you would get a strange combination of a triangle wave and a sawtooth wave.

Modified sawtooth-wave generator

Many standard signal-generator circuits require that certain component values be balanced to achieve the best possible waveform. If the required component values don't have the correct equalities, the output waveform will be distorted. Some very unusual waveforms can be created by selectively distorting a standard waveform.

For example, the circuit shown in Fig. 6-11 is a sawtooth-wave signal generator. Ordinarily, capacitors C1 and C2 must have equal values. The output will then be a pretty good ascending sawtooth wave, as illustrated in Fig. 6-12.

We can deliberately distort the output waveform from this circuit by changing the value of one of the capacitors. Figure 6-13 shows a simple modification of the ba-

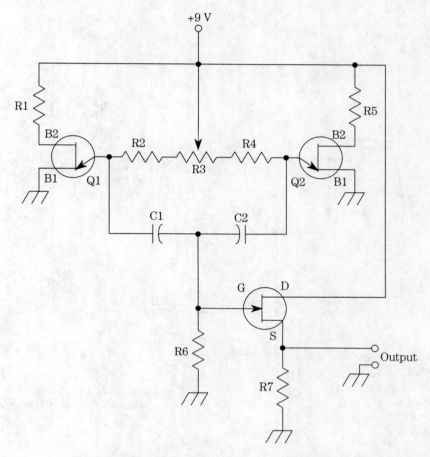

6-11 This circuit is a sawtooth-wave signal generator.

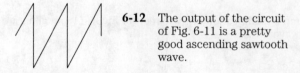

6-12 The output of the circuit of Fig. 6-11 is a pretty good ascending sawtooth wave.

sic circuit that permits it to generate six different waveforms. One is a regular saw-tooth wave, while the other five are unnamed and difficult to describe. A couple of typical output waveforms from this circuit are illustrated in Fig. 6-14. These bizarre waveforms have very different harmonic make-ups than a standard sawtooth wave. They can also produce some very unusual effects if they are used as control signals.

This circuit is easy to use. Each waveshape is switch-selectable by a 6-position rotary switch. The output signal frequency can be fine-tuned via potentiometer R3. Unfortunately, there is no practical frequency equation for this circuit, because the capacitors are part of what determines the cycle timing. Therefore, changing the

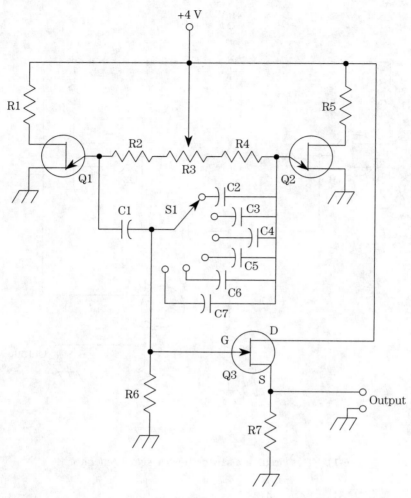

6-13 This simple modification of the circuit of Fig. 6-11 can generate six different waveforms.

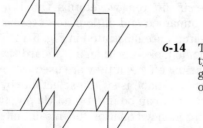

6-14 These are a couple of typical output waveforms generated by the circuit of Fig. 6-13.

waveshape with the switch alters the signal frequency. Still, this is a nice circuit for experimentation. A suggested parts list for this modified sawtooth-wave generator circuit is given in Table 6-2.

Table 6-2. Suggested parts list for the
modified sawtooth-wave generator circuit of Fig. 6-13

Q1, Q2	UJT (TIS-43, or similar)
Q3	FET (MPF103, Radio Shack RS2028, or similar)
C1, C4	0.01-μF capacitor
C2	0.001-μF capacitor
C3	0.0047-μF capacitor
C5	0.022-μF capacitor
C6	0.047-μF capacitor
C7	0.1-μF capacitor
R1, R5, R7	4.7-kΩ, ¼-W, 5% resistor
R2, R4	100-kΩ, ¼-W, 5% resistor
R3	1-MΩ potentiometer
R6	1-MΩ, ¼-W, 5% resistor
S1	SP6T rotary switch

Staircase-wave generators

A fairly popular non-standard waveform is the staircase wave. It is derived from a rectangle wave (usually, though not always, a square wave). There are many different types of staircase waves. They can be classified by the number and direction of the steps. Figure 6-15 shows an ascending staircase wave. A descending staircase wave is shown in Fig. 6-16. Some staircase waves go both up and down, as illustrated in Fig. 6-17. As you can see, all of these waveforms look rather like staircases, hence the name.

A very simple circuit for converting a square-wave input into a staircase-wave output is shown in Fig. 6-18. This is not a practical circuit. It is a conceptual demonstration only. Its output signal is in the form of an ascending staircase wave. Each

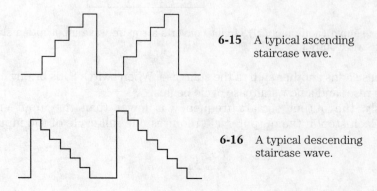

6-15 A typical ascending staircase wave.

6-16 A typical descending staircase wave.

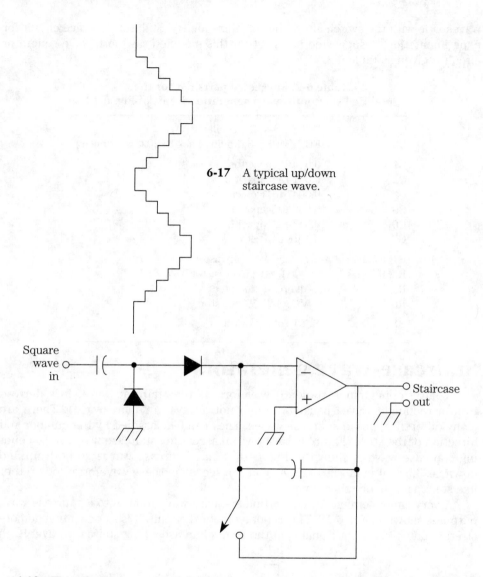

6-17   A typical up/down
staircase wave.

6-18   This simple (non-practical) circuit converts a square-wave input into a staircase-wave output.

new input pulse adds another step to the staircase. When switch S1 is briefly closed,
the circuit is reset and a new staircase cycle begins.

Obviously, the output signal's frequency is lower than the input signal's
frequency. Each step in the output cycle requires one full cycle of the input signal, so:

$$F_o = \frac{F_i}{N}$$

where F_o is the output frequency, F_i is the input frequency, and N is the number of steps in the output staircase signal. We are assuming that there are an equal number of steps in each complete output cycle.

It doesn't take much deep thought to realize that a manual switch to force-start each new cycle would not be practical in most applications. We need to automate the process to get a useful staircase-wave generator circuit.

Many different types of electronic switching can be used in such an application. The block diagram of Fig. 6-19 illustrates how a second square-wave generator and a digital gate can be used in place of the mechanical switch. The control signal generator (the second square-wave generator) resets the staircase on each of its own positive half-cycles. Naturally, the control signal generator must have a much lower frequency than the input square-wave generator. The staircase wave at the output will have a frequency equal to that of the control generator. The ratio of the control and input frequencies will determine the number of steps in each staircase cycle:

$$N = \frac{F_i}{F_c}$$

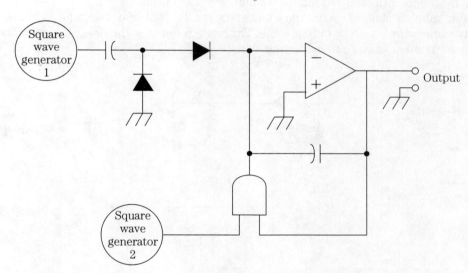

6-19　A second square-wave generator and a digital gate can be used in place of the mechanical switch is a staircase-wave generator circuit.

where F_c is the frequency of the control signal (from the second square-wave generator), F_i is the input signal frequency (from the first square-wave generator), and N is the number of steps in the output staircase signal.

For the best results, the two signal generators in this circuit should have frequencies that are exact multiples of each other. This will ensure that each cycle of the staircase wave at the output will always have the same number of steps. For example, let's say the control signal frequency (F_c) is 100 Hz, and the input signal frequency (F_i) is 500 Hz. The output signal will be a staircase wave with a frequency of 100 Hz, and five steps per cycle:

$$N = \frac{500}{100}$$

$$= 5$$

However, if we change the input signal frequency (F_i) to 475 Hz, but keep the control signal frequency (F_c) at 100 Hz, we will not get a nice, even number of steps-per-cycle;

$$N = \frac{475}{100}$$

$$= 4.75$$

On some cycles, the output signal will have four steps, but on other cycles it will have five steps—depending on the present (momentary) phase relationship between the two square-wave signals at that particular instant. Because their frequencies are not harmonically related, their relative phase will be constantly changing.

Another, somewhat more sophisticated approach is illustrated in Fig. 6-20. In this system we are using just a single square-wave generator, so there will never be any shifting phase problems. The control signal is derived directly from the input signal by counting the desired number of input pulses (output steps). After a specific, preset number of input cycles, the counter triggers the gate and resets the staircase-wave generator for a new output cycle. This system removes the possibility of an irregular number of steps in the output signal.

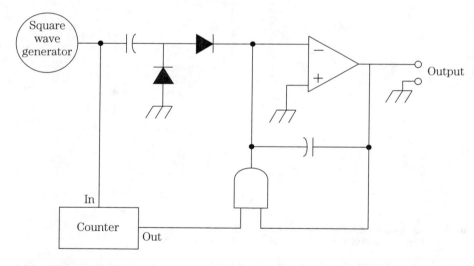

6-20   This circuit demonstrates a somewhat more sophisticated approach to making a practical staircase-wave generator.

In any practical staircase generator circuit, the maximum number of output steps will be limited by the op amp's supply voltage. An op amp's output voltage can never exceed its supply voltage. If we try to force it to put out a higher voltage, the output signal will be clipped, as illustrated in Fig. 6-21.

The lower the amplitude of the input signal, the greater the number of steps you will be able to fit into the output signal.

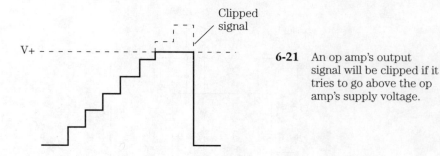

6-21 An op amp's output signal will be clipped if it tries to go above the op amp's supply voltage.

Project #9—Up/down staircase-wave generator

The staircase-wave generator circuit we will be using in this project uses a somewhat different approach than those discussed above. The output signal from this circuit is an up/down staircase wave. It builds up from zero to a specific peak, then the steps go downwards back to zero before the next cycle begins.

The circuit we will be working with is shown in Fig. 6-22. A suitable parts list for this project is given in Table 6-3. Nothing is too critical, so experimenting with alternate component values on a breadboard is inviting.

Adjusting potentiometer R1 controls the frequency. To change the entire frequency range, you can try substituting different values for capacitor C1.

IC1 is a 7555 timer. This is a CMOS version of the popular 555. If you have trouble finding the 7555, you can substitute a 555 chip. This substitution won't hurt anything, and will not require any supporting changes in the rest of the circuitry at all. The 555 and the 7555 are electrically identical, as far as their functioning goes. They are even pin-for-pin compatible, so you don't even have to worry about correcting the pin numbers. The biggest functional difference is that the 7555 consumes somewhat less current than the 555.

If a fairly large filter capacitor is added between the output line and ground of this circuit, the output signal will be a fair digital approximation of a triangle wave, as illustrated in Fig. 6-23. This is accomplished simply by the filter capacitor smoothing out the sharp corners of the staircase steps.

Project #10— 555 unusual waveform generator

By using one signal generator to control a second, we can generate a wide variety of unusual waveforms. A very simple but versatile unusual waveform generator circuit is shown in Fig. 6-24. Here we are using two 555 timers to generate many different pulse type signals.

A suitable parts list for this project is given in Table 6-4. Nothing is critical in this circuit, and you are encouraged to experiment with alternate component values throughout the circuit. Capacitors C13 and C14 are included only to ensure stability

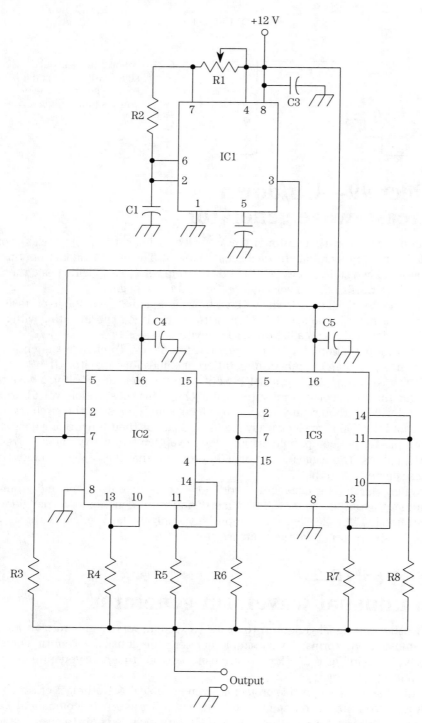

6-22 Project #9—Up/down staircase-wave generator.

**Table 6-3. Suggested parts
list for Project #9—Up/down
staircase-wave generator of Fig. 6-22**

IC1	7555 timer (or 555, see text)
IC2, IC3	CD4042 quad latch
C1	0.1-μF capacitor
C2	0.01-μF capacitor
R1	100-kΩ potentiometer
R2	10-kΩ, ¼-W, 5% resistor
R3, R5	22-kΩ, ¼-W, 5% resistor
R4, R7	33-kΩ, ¼-W, 5% resistor
R6, R8	27-kΩ, ¼-W, 5% resistor

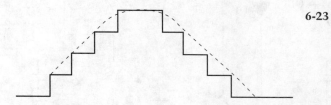

6-23 If a fairly large filter capacitor is placed across the output of the circuit of Fig. 6-22, the output signal will be a fair digital approximation of a triangle wave.

of the timer chips. The exact value of these capacitors has no noticeable effect on the operation of this circuit, or on its output signals.

This is a pretty simple circuit. IC1 is an astable multivibrator or rectangle wave generator circuit. Its signal frequency and duty cycle are determined by the settings of potentiometers R1 and R3, along with the capacitor value selected via switch S1.

The output signal (pin #3) from IC1 is fed into the trigger input (pin #2) of IC2, which is wired as a monostable multivibrator circuit. Its timing period is determined by the setting of potentiometer R5 and the capacitor value selected by switch S2.

The outputs of both timer stages are combined through D1, R7 and R8 before being fed out to the circuit output. You might want to replace resistor R7 with a 10-kΩ or 25-kΩ potentiometer, for even more possible effects. The diode prevents the output of IC2 from being fed back into its own trigger input, avoiding the possibility of a latch-up condition.

By twiddling with the three potentiometers, and trying various settings of the two rotary switches, you can produce a wide variety of very, unusual tones. Some will sound very pleasant, and even musical when heard through an audio amplifier and loudspeaker. Others will be extremely obnoxious.

Some combinations of control settings in this project could result in no output tone at all. If this happens, try readjusting some or all of the controls before concluding that anything is wrong.

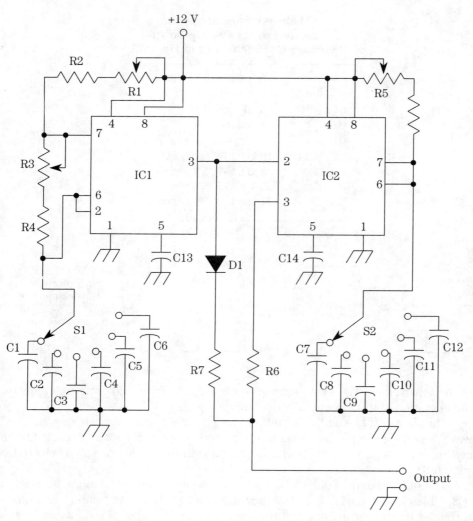

6-24   Project #10—555 unusual waveform generator.

**Table 6-4. Suggested parts list for Project #10—
555 unusual waveform generator of Fig. 6-24**

IC1, IC2	555 timer
D1	Signal diode (1N4148, 1N914, or similar)
C1, C7, C13, C14	0.01-µF capacitor
C2, C8	0.047-µF capacitor
C3, C8	0.1-µF capacitor
C4, C9	0.47-µF capacitor
C5, C11	1-µF, 25-V electrolytic capacitor
C6, C12	5-µF, 25-V electrolytic capacitor

R1, R3, R5	100-kΩ potentiometer
R2, R4, R6, R8	1-kΩ, ¼-W, 5% resistor
R7	4.7-kΩ, ¼-W, 5% resistor
S1, S2	SP6T rotary switch

Project #11—Programmable waveform generator

An extremely versatile unusual waveform generator circuit is shown in Fig. 6-25. This one is digitally-based, and is programmable. This gadget uses ten CMOS bilateral switches (IC3, IC4, and IC5) and ten potentiometers (R3 through R12) to permit literally thousands of differing waveforms to be programmed. A suitable parts list for this project is given in Table 6-5.

In operation, this circuit goes through a ten-step sequence, sampling the voltage across each of the programming potentiometers one at a time. The bilateral switches select one of the potentiometers for each step in the sequence counted out by IC2.

The sequence rate is determined by the frequency of the astable multivibrator (rectangle-wave generator) built around IC1, a standard 555 timer. One each output pulse from this 555, the counter (IC2) advances one step in the sequence. When the sequence is completed, the counter automatically resets itself and starts a new sequence from zero.

The sequence step frequency can be adjusted via potentiometer R1. You might also want to experiment with different values for capacitor C1. If this capacitor is made larger than 1 µF to 5 µF, and you can use this project as a ten-step voltage sequencer.

By keeping the 555's frequency well up into the audible range, the output will be a complex audio frequency waveform, with lots of sidebands. Basically, this circuit is a sort of staircase-wave generator—except each successive step can move any distance in either direction. A typical output signal from this circuit is illustrated in Fig. 6-26. The height of each step is determined by the setting of the potentiometer that is activated at that point in the sequence. The ten-step potentiometers (R3 through R12) interact to create each programmed waveform.

To minimize the project's cost and bulk, you might want to consider using miniature trimpots for R3 through R12. However, this will make the circuit a little less convenient to use. You will need to use a screwdriver to program a new output waveshape. For maximum convenience, ten standard front-panel potentiometers would probably be the best bet in this circuit. This will leave a lot of empty space behind the control panel, because the rest of the circuitry is so compact. While this might seem a little inelegant, there is certainly no harm in it.

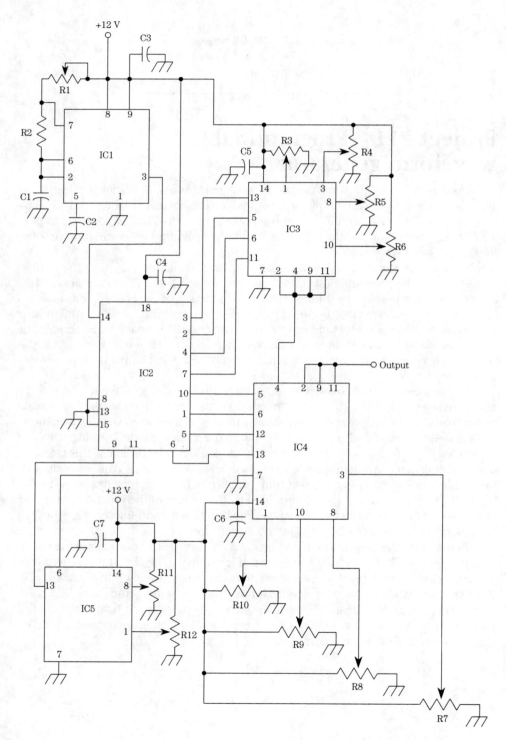

6-25 Project #11—Programmable waveform generator.

**Table 6-5. Suggested parts list for Project #11—
Programmable waveform generator of Fig. 6-25**

IC1	7555 timer (or 555)
IC2	CD4017 decade counter
IC3, IC4, IC5	CD4066 quad bilateral switch
C1	0.1-μF capacitor (see text)
C2	0.01-μF capacitor
R1	500-Ω potentiometer (frequency)
R2	1-kΩ, ¼-W, 5% resistor
R3–R12	10-kΩ potentiometer (step values) (see text)

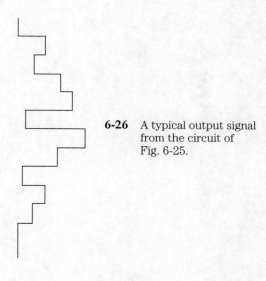

6-26 A typical output signal
from the circuit of
Fig. 6-25.

7
CHAPTER

RF oscillators and signal generators

In principle, RF (radio frequency) oscillators and signal generators aren't all that dissimilar to AF audio frequency circuits—except for their much higher operating frequencies. An AF oscillator nominally operates in the 20 Hz to 20 kHz (20,000 Hz) range. An RF oscillator typically runs at least several hundred kilohertz (1 kHz = 1,000 Hz), and often well into the Megahertz (1 MHz = 1,000 kHz = 1,000,000 Hz) range. This much higher frequency range puts some added constraints on the circuit design, and because of that, a number of special problems can crop up.

Most RF oscillators are true sine-wave oscillators. This is because RF signals are usually modulated, most commonly by amplitude modulation or frequency modulation, as discussed in chapter 6. The program signal is usually a complex audio signal such as speech and/or music. As we learned earlier, modulation generally works best if at least one of the input signals is a sine wave, or else the sideband generation can really get out of hand, resulting in an incredibly muddy, essentially useless mess of an overly-complex signal. This is why RF carrier signals are almost always sine waves.

Most of the discrete sine-wave oscillator circuits of chapter 2 can be adapted for RF work. Crystal oscillators almost always operate in the RF region.

Stray capacitances

As frequency increases, so does the sensitivity to capacitance. Capacitances that are totally negligible at audio frequencies can become exceedingly significant at radio frequencies.

A dc resistor functions pretty much the same way, regardless of the signal frequency, but inductors and capacitors have ac resistance components, called *reac-*

tance. Capacitive reactance decreases with increases in the signal frequency. The formula for determining capacitive reactance is:

$$X_c = \frac{1}{(2\,\pi\,FC)}$$

where F is the signal frequency in Hertz, C is the capacitance in farads, and X_c is the capacitive reactance in ohms. π, or pi, is a standard mathematical constant, that always has a value of approximately 3.14, so the formula can be re-written as:

$$X_c = \frac{1}{(6.28FC)}$$

Let's assume we have a 10-μF (0.00001 farad) capacitor. If the signal frequency is 150 Hz, the capacitive reactance is:

$$X_c = \frac{1}{(6.28 \times 150 \times 0.00001)}$$

$$= \frac{1}{0.00942}$$

$$= 106 \text{ ohms}$$

But if we increase the signal frequency applied across that same 10-μF capacitor, to 2.2 kHz (2200 Hz), the capacitive reactance becomes:

$$X_c = \frac{1}{(6.28 \times 2200 \times 0.00001)}$$

$$= \frac{1}{0.13816}$$

$$= 7.24 \text{ ohms}$$

Quite an appreciable change. At radio frequencies, this 10 μF becomes almost invisible as a reactance. For example, at 1760 kHz (1,760,000 Hz), the capacitive reactance is only:

$$X_c = \frac{1}{(6.28 \times 1760000 \times 0.00001)}$$

$$= \frac{1}{110.528}$$

$$= 0.01 \text{ ohm}$$

The larger the capacitance, the lower the reactance at a given signal frequency. We will see the significance of this shortly.

A capacitor is ultimately nothing more than two conductors (called plates), separated by an *insulator* (called the *dielectric*). The plates can be any two wires, and ordinary air can serve as a dielectric. Any two (or more) closely spaced wires will electrically act as if there is a *phantom capacitor* connected between them, as illustrated in Fig. 7-1. Adjacent traces on a printed circuit board form even better phantom capacitors.

In most circuits, including audio frequency circuits, such phantom capacitors are of little or no importance. They usually won't affect the operation of the circuit

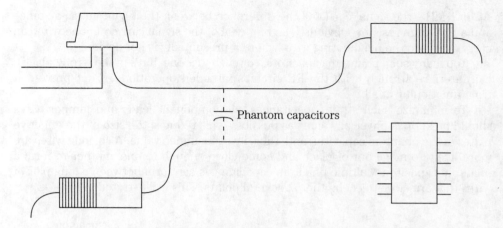

7-1 Any two closely-spaced wires will electrically act as if a phantom capacitor is connected between them.

in any noticeable way. This is because the capacitance is so small and the reactance is so high. Typically, a stray capacitance will only have a value of a few dozen picofarads at most. One microfarad, which is one millionth of a farad, is equal to one million picofarads. We are talking about a very small value here. That is:

$$1 \text{ pF} = 0.000001 \text{ } \mu\text{F}$$

$$= 0.000000000001 \text{ farad}$$

Let's assume a fairly typical stray capacitance value, let's say 25 pF. Actually this is probably a bit large.

If the signal frequency encountering this phantom capacitor is 250 Hz, the capacitive reactance will work out to:

$$X_c = \frac{1}{(6.28 \times 250 \times 0.000000000025)}$$

$$= \frac{1}{0.00000003925}$$

$$= 25,500,000 \text{ ohms}$$

In other words, there is very good insulation between the two wires. No noticeable amount of the signal carried in one wire will jump over to its neighbor.

But when we move up to radio frequencies, even this tiny phantom capacitor is likely to become significant. For example, at a signal frequency of 89.1 MHz (89,100,000 Hz), the capacitive reactance works out to:

$$X_c = \frac{1}{(6.28 \times 89100000 \times 0.000000000025)}$$

$$= \frac{1}{0.0139887}$$

$$= 71.5 \text{ ohms}$$

At 89.1 MHz, the signal can't tell the difference between the phantom capacitance and a 71.5-ohm resistor. Obviously, a great deal of the signal carried in one wire can carry over into the neighboring wire, where it presumedly is not desired.

Too many such phantom capacitors, (or even just one, if it's in the wrong place), can interfere strongly with the RF circuit's intended operation—if not prevent it from functioning at all.

To minimize such stray capacitances, all component leads and jumper wires should be kept as physically short as possible. This reduces the size of the effective plates of the phantom capacitor, and, thus, its capacitance value. Also, long wires are more likely to move out of place and come close enough to one another to form a phantom capacitor. Multiple phantom capacitances can form between a long pair of wires that are close to each other at several points. This is illustrated in Fig. 7-2.

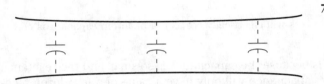

7-2 Multiple phantom capacitances can form between a long pair of wires that are close to each other at several points.

For wires that must run more than a couple of inches, shielding is recommended. That is, a grounded (usually, though not always) conductor is placed around the actual conductor. Coaxial cable is often used, as shown in Fig. 7-3. Besides blocking off the formation of any phantom capacitors with other conductors, this shielding will also help limit interference from RF pickup. Any uninsulated wire can act as an antenna, turning almost any electronic component into a sort of mini-radio receiver. Again, this puts undesired signals where they don't belong in the circuit.

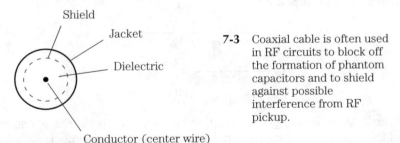

7-3 Coaxial cable is often used in RF circuits to block off the formation of phantom capacitors and to shield against possible interference from RF pickup.

A coaxial or shielded cable is not always the best, or even an adequate solution against the problem of phantom capacitors, however. Remember, the shielding is also a conductor, so it can act like a capacitor plate, and the insulation between the actual central conductor and the shield can act like a dielectric. In fact, in a coaxial cable, this insulation is officially called the *dielectric*. This means if the shield is grounded, there might be one or more phantom capacitors between the intended conductor and ground. In most cases this won't do any harm, besides wasting some power.

A signal shunted to ground generally doesn't accomplish much, other than being converted into a little waste heat. But in some circuits, too much of the desired signal could be inadvertently shunted to ground through one or more phantom capacitors. This inevitably detracts from the overall signal strength. In some cases, there may not be enough of the desired signal left for the circuit to operate reliably, or perhaps even at all.

In some cases, it might be best to leave the shield ungrounded. You might want to connect it to something else in the circuit. In some applications, it could be desirable to leave the shield unconnected electrically to anything, other than itself.

Adjacent printed circuit board traces can be effectively shielded against forming phantom capacitances by separating them with a third *guard band* trace, as illustrated in Fig. 7-4. In most cases, the guard band will be grounded, but this is subject to the same occasional limitations described above for shielded cables.

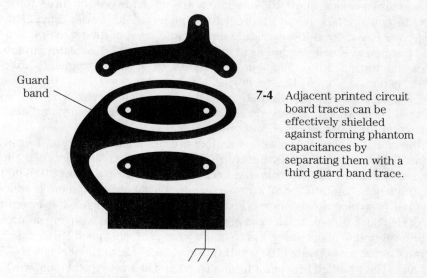

Guard band

7-4 Adjacent printed circuit board traces can be effectively shielded against forming phantom capacitances by separating them with a third guard band trace.

Stray capacitances can also form within certain electronic components, especially semiconductors. For example, a typical bipolar transistor looks electrically like Fig. 7-5 in a high-frequency circuit. These internal capacitances are typically very tiny, and are totally negligible in audio frequency applications. But at radio frequencies, even the smallest stray capacitance can make a big difference.

There isn't much the circuit designer or electronics hobbyist can do about such internal capacitances, because they are built into the component itself. All that can be done is to make sure to use components rated by their manufacturers for use in

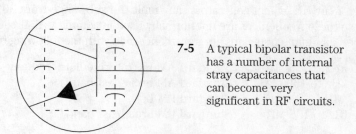

7-5 A typical bipolar transistor has a number of internal stray capacitances that can become very significant in RF circuits.

circuits involving the desired frequency (or frequencies). Specialized semiconductor components have been designed expressly for use in RF circuits.

Internal capacitances are one of the primary factors in defining a components rated operating frequency specification. A transistor designed for use in audio amplifiers and similar applications isn't likely to be of much use in RF circuits. A transistor designed for RF use can probably be used in most audio frequency applications, but there would be little reason to do so. An RF transistor is almost certain to be a lot more expensive than an AF transistor with similar power-handling specifications.

RF signal testing

Generally speaking, most RF oscillators are used for one of three purposes. First, radio transmitters use an RF oscillator to generate the carrier signal. Most radio receivers have a local RF oscillator circuit that is used in tuning the radio to the desired frequency. Finally, and of most interest to us here, RF oscillator circuits are often used in testing radio equipment. In this case, the device is frequently called an RF signal generator. Some older technical literature may also refer to it as a test oscillator, but this usage is rather rare today. The output signal from an RF signal generator could be a straight, unmodulated sine wave at an RF frequency, or it might be modulated in some very specific, and predictable way.

Professional-quality RF signal generators are designed to operate over a very wide frequency range, nominally the entire RF spectrum. In recent decades, the upper end of the usable RF spectrum has been moved increasingly upward. When the signal frequencies get into the microwave region, specialized equipment is generally called for. Everything is extremely critical at such high frequencies. It is beyond the scope of this book to look at microwave compatible equipment with any depth.

For most practical purposes, the frequency range of most good RF signal generators run from about 100 kHz (100,000 Hz) to about 120 MHz (120,000,000 Hz). This covers all of the standard broadcast frequencies used in most typical radio work. Because the frequency range is so wide, it is usually broken up into over-lapping sub-ranges, that can be selected via a multi-position switch. For example, a typical RF signal generator might have five sub-ranges:

100 kHz	–	600 kHz
500 kHz	–	2,200 kHz (2.2 MHz)
2 MHz	–	25 MHz
20 MHz	–	90 MHz
90 MHz	–	120 MHz

Actually, many RF signal generators may omit the sub-range from 20 MHz to 80 MHz, because these frequencies are not normally used in commercial, broadcast radio. Most commercial radio technicians are most concerned with the following frequencies:

455 kHz	(Standard AM IF (intermediate frequency))
560 kHz to 1600 kHz	(Standard AM broadcast band)
10.7 MHz	(Standard FM IF)
88 MHz to 108 MHz	(Standard FM broadcast band)

Of course, if a technician intends to work on other types of radios—such as shortwave, ham, or CB equipment, they will need an RF signal generator that can supply signals at the appropriate frequencies.

Sometimes an RF signal generator is used in servicing a television set, but this won't often be very practical. Special video signal generators are available for this purpose. Such devices are beyond the scope of this book.

Most standard RF signal generators also include an audio-frequency oscillator stage. Usually, the signal frequency of this audio oscillator is not adjustable. Typically it is fixed at 1,000 Hz (1 kHz), a good, mid-range frequency. This audio signal can usually be accessed directly through the RF signal generator's output, if needed, but its primary purpose is as the program signal when a modulated RF signal is needed. A switch selects between a modulated or unmodulated RF output signal, and sometimes the fixed AF tone as well. In most cases, amplitude modulation (AM) is used in RF signal generators. Some deluxe units also offer frequency modulation (FM), or perhaps other less common types of modulation as well.

The chief uses of the unmodulated RF signal include substituting it for a questionable local oscillator circuit in a receiver under test. In some tests, it can be useful to adjust the unmodulated RF test signal to beat with other signal frequencies being carried by the circuit being tested.

Usually, the modulated RF test signal will be used more by the average technician than the unmodulated RF signal. It is used primarily as a substitute for an over-the-air signal, so there is no confusion resulting from an unpredictable and/or unreliable broadcast signal. With the RF signal generator, you always know exactly what the modulating program signal is at all times—a continuous 1-kHz tone. There is no need to worry about an antenna or the received signal strength, or possible interference signals. If the radio station's signal happens to be very quiet or even silent for a few seconds—such as if a DJ doesn't catch the end of a record quickly enough—it could cause confusion if you are adjusting something by listening to the demodulated program signal. Normal broadcast programs go up and down in amplitude unpredictably, making certain adjustments difficult.

By tuning the RF signal generator for the receiver's intermediate frequency (IF), the technician can bypass the receiver's RF input stages, and inject the test signal into a later stage of the circuitry.

Another control found on all commercial-grade RF signal generators is the attenuator. This sets the amplitude or strength of the output signal.

A step attenuator is a multi-position switch with several discrete attenuation positions, such as:

X0.1
X1
X10
X100
X1000

and so forth. The higher the multiplier value, the greater the amplitude of the output signal.

A continuous attenuator is a potentiometer that acts pretty much like a volume control on an audio amplifier. Some RF signal generators combine both a step attenuator and a continuous attenuator, for maximum control and accuracy. The step attenuator is first set to the desired range, then the continuous attenuator is adjusted to fine-tune the output signal level.

The attenuator controls are usually calibrated in millivolts (1 millivolt = 0.001 volt). If the test signal's amplitude is too high, the circuit under test might overload, and the results obtained from the test will be thrown off, and could be totally invalid and meaningless. On the other hand, if the test signal's level is too weak, the circuit under test may not be able to function properly either. The exact signal level usually isn't too terribly critical, but in most cases it should be reasonably close.

In particularly critical test procedures, it's best not to trust the calibration markings on an RF signal generator's attenuator controls too exactly. When in doubt, check the actual output signal level with a good ac voltmeter, or use an oscilloscope.

The basic RF signal generator as discussed so far in this section is adequate for most normal testing of AM receivers. Testing FM receivers is generally a bit more complicated. An RF sweep-signal generator is often the best choice for FM work. Remember that in an FM signal, the carrier frequency is deviated from its nominal value by an amount proportional to the program signal. The same approximate effect, for testing purposes, can be achieved by sweeping the RF test signal through a specific range, corresponding to the maximum deviation of the FM signal it is intended to simulate. In other words, a VCO (voltage-controlled oscillator) is used to generate the RF test signal, while the control voltage is an audio frequency, fixed-program signal.

The amount of modulation can be adjusted by controlling the amplitude of the audio frequency-modulation signal. At its minimum amplitude (zero), the output signal is unmodulated. The full range setting on most commercial RF sweep-signal generators creates a maximum deviation of 450 kHz (or ±225 kHz). For example, if the nominal RF signal frequency is set for 2 MHz (2000 kHz), the actual signal will fluctuate between 1.775 MHz and 2.225 MHz.

Of course, in practical usage for testing broadcast FM receivers, the RF test signal frequency (carrier) will be set either at the standard IF (intermediate frequency) of 10.7 MHz, or somewhere within the standard FM broadcast band (88 MHz to 108 MHz).

The RF sweep-signal generator is basically similar in concept to the audio frequency sweep-signal generator project that was presented back in chapter 4 of this book. The actual circuitry will tend to be somewhat more complex, however, in order to accommodate the much higher frequencies required in FM work. The modulation depth (amplitude of the program signal) must also be more precisely controllable in an RF sweep-signal generator.

Inside an RF signal generator

A simple, but effective RF signal generator circuit is shown in Fig. 7-6. FET Q1 is the heart of the actual RF oscillator. This is a Hartley oscillator, described back in chapter 2. This is one of the simplest and most common types of oscillator circuits. Of course, its output signal is in the form of a sine wave.

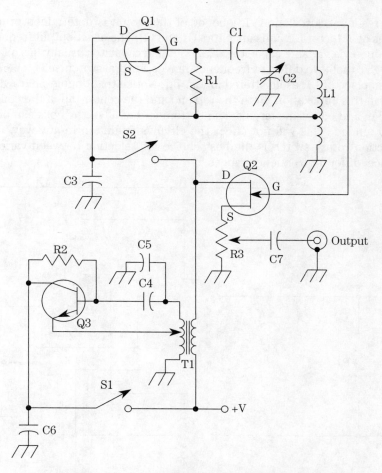

7-6   This is a simple, but effective RF signal-generator circuit.

The Hartley oscillator circuit is often referred to as a *split-inductance oscillator* because coil L1 is center-tapped, making it a sort of autotransformer. In effect, L1 acts as if it was two separate coils in very close physical proximity, with a common connection point (the center-tap). For convenience, we will call the topmost terminal of L1 "A," the center-tap "B," and the bottom-most (grounded) terminal of the coil "C." A current through coil section AB induces a proportional signal into coil section BC, and vice-versa.

A FET is used in this circuit instead of a more common bipolar transistor (as in chapter 2), because bipolar transistors generally don't work as well at high RF frequencies. The natural high input impedance and low output impedance of a FET is much better suited to such high frequency applications. Otherwise, this Hartley oscillator is not significantly dissimilar to the basic AF Hartley oscillator circuits introduced back in chapter 2.

The resonant (oscillation) frequency of the circuit is determined primarily by the values of inductor L1, and capacitors C1 and C2. The exact output frequency can be fine-tuned via variable capacitor C2. A variable capacitor usually has a fairly narrow range. To achieve different frequency ranges, several switch-selectable coils can be used in place of L1, as illustrated in Fig. 7-7. Notice that double-pole switching is required in this application. Both the top end and the center-tap of the coil must be switched in and out of the circuit. The bottom-end of all of the coils can be permanently grounded. This will not affect the circuit's operation in any way. Usually a multi-section rotary switch is the best choice for selecting between various coils, and, hence, different frequency ranges.

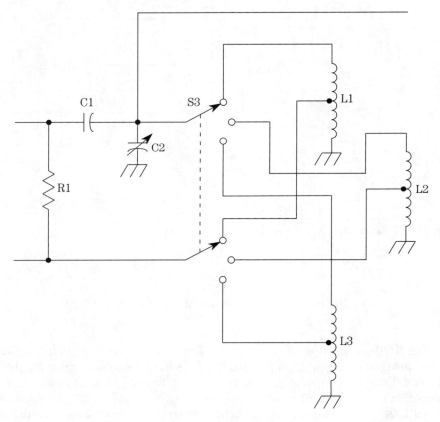

7-7 To achieve different frequency ranges in the circuit of Fig. 7-6, several switch-selectable coils can be used in place of L1.

The second FET is wired as a source follower, or a simple buffer amplifier stage, to prevent the effects of possible loading from the output circuit or device being fed the RF output signal. The output signal level, or amplitude, can be adjusted via potentiometer R3. Capacitor C3 blocks any possible dc component in the output signal, preventing it from damaging or confusing load circuits or devices that are driven by this RF signal generator.

A second Hartley oscillator circuit is constructed around transistor Q3. Because this oscillator is designed to generate a much lower (AF) signal frequency, a less expensive bipolar transistor can be used. There is no need to use a FET in this stage. This oscillator has a frequency in the audible range, and it is not variable. A typical frequency would be about 1,000 Hz (1 kHz), although this might vary somewhat, depending mainly upon the personal preferences of the individual circuit designer.

While the signal frequency generated by this stage is fixed, the audio oscillator itself is switchable. When switch S1 is open, the audio oscillator circuit gets no power, so, of course, it has no output signal. The final signal output from the entire signal generator will simply be the unmodulated RF signal, generated by Q1.

Closing switch S1, however, turns on the audio oscillator circuit, and its output signal amplitude-modulates the RF signal (from Q1) as they both pass through the output amplifier (Q2). Similarly, switch S2 controls the supply voltage to the RF oscillator stage (Q1). When this switch is open, and switch S1 is closed, the straight AF tone (generated by Q3) will appear at the output.

What happens if both of these switches are opened? Nothing at all, actually. There will be no output signal from this circuit until one or both of these switches are closed. To make things even clearer, the functional elements of this RF signal-generator circuit are broken down into simple block diagram form in Fig. 7-8.

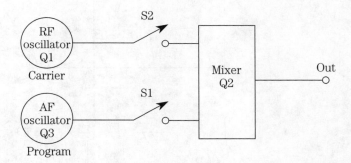

7-8 This block diagram shows the functional elements of the RF signal-generator circuit of Fig. 7-6.

Because of the high (RF) frequencies involved, this circuit must be well-shielded throughout. All component leads and connecting wires should be as physically short as possible. The output cable connecting the RF signal generator to the output load circuit or device should also be shielded, and as short as possible to minimize the possibility of serious interference problems.

Project #12—
Unmodulated RF signal generator

A practical RF signal-generator circuit is illustrated in Fig. 7-9. A suitable parts list for this project is given in Table 7-1. Using the component values suggested in the

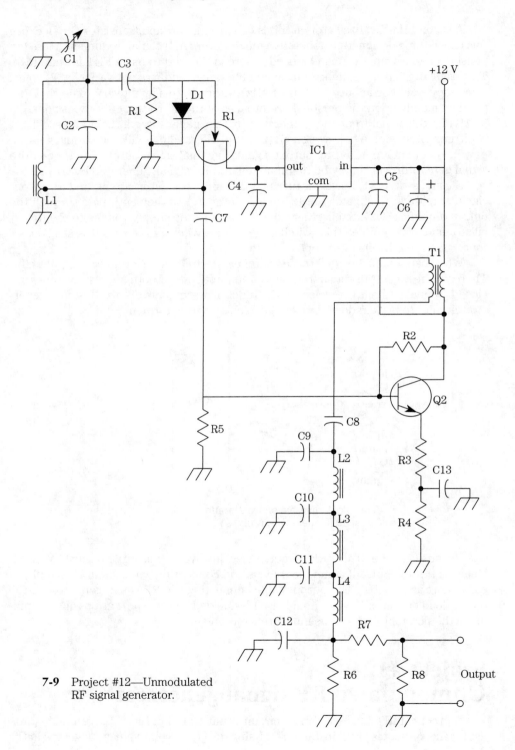

7-9 Project #12—Unmodulated
RF signal generator.

parts list, the output signal frequency will be about 3.7 MHz, and the output signal strength should be about 10 milliwatts into a 50-ohm load.

Table 7-1. Suggested parts list for
Project #12—Unmodulated RF signal generator of Fig. 7-9

IC1	7805 +5 V voltage regulator
Q1	FET (Radio Shack RS2035, 2N5486, MPF-102, or similar)
Q2	npn transistor (2N5109, ECG128, Radio Shack RS2030, or similar)
C1	Air trimmer capacitor—1.4 pF to 9.2 pF
C2	300-pF silvered-mica capacitor
C3	5-pF capacitor
C4, C5	0.01-μF capacitor
C6	220-μF, 25-V electrolytic capacitor
C7	56-pF silvered-mica capacitor
C8, C13	0.1-μF capacitor
C9, C12	470-pF silvered-mica capacitor
C10, C11	1200-pF silvered-mica capacitor
T1	7 turns of #22 enamelled wire bifilar-wound core = Amidon FT-37-43, or similar
L1	7.5-μH coil center-tap 2-μH
L2, L$	2.5-μH coil
L3	3.0-μH coil
R1	1-MΩ, ¼-W, 5% resistor
R2	1-kΩ, ¼-W, 5% resistor
R3	4.7-Ω, ¼-W, 5% resistor
R4	47-Ω, ¼-W, 5% resistor
R5	470-Ω, ¼-W, 5% resistor
R6, R8	150-Ω, ¼-W, 5% resistor
R7	39-Ω, ¼-W, 5% resistor

Some of these component's values are somewhat critical. Do not use ordinary low-grade ceramic disc capacitors for those components the parts list identifies as silvered-mica capacitors. If you cannot find silvered-mica devices, substitute a similar capacitor type with an NP0 temperature coefficient.

FET Q1 and its associated components form an RF Hartley oscillator circuit. An MPF102 FET should work OK in this circuit, but the RF signal strength at the project's output may be a little lower if this device is used. However, it will probably be easier to find, and less expensive than the 2N5486. Will the reduced signal strength matter? That depends on the specific requirements of the particular application you have in mind for this project. In a pinch, you could always feed the output from this RF signal generator through a suitable RF amplifier circuit to boost the overall signal level to the desired amplitude.

Capacitor C1 is used to fine-tune the RF signal frequency. This device's specifications are not very critical. You should be able to use almost any small trimmer capacitor cannibalized from some other, discarded piece of radio equipment. If it's range is very large, the RF signal generator will be very difficult to tune precisely.

For more flexibility in the RF signal frequency, you can substitute different coil values for L1. You could use a double-pole rotary switch to select between several different coils, and therefore, frequency ranges, as discussed in the preceding section.

Another way to change the RF signal frequency in this circuit is to change the value of capacitor C2. Increasing this capacitance will decrease the signal frequency, and vice versa. You can use one or more switches to add one or more additional capacitors in parallel with C2. Remember, capacitances in parallel add their values. To keep the signal in the RF range, the total capacitance value should remain fairly small.

Transistor Q2 is the heart of a class A buffer amplifier. The RF output signal is not modulated in this project. That would not be a difficult modification to make. Just add a suitable AF (audio frequency) oscillator circuit that controls the gain of an RF amplifier. This circuit has been designed to give maximum performance at minimum cost. If you have difficulty finding the coils, you can wind your own. Coil L1 is 31 turns of #18 enamelled wire, wound on an Amidon T-94-6 core. The center-tap connection is made 8 turns above the grounded terminal. The remaining coils (L2, L3 and L4) all use #22 enamelled wire on an Amidon T-50-2 core. Use a separate core for each coil. L2 and L4 should have 21 turns apiece, and L3 should have 23 turns.

The oscillator circuit and buffer amplifier in this circuit are a little crude, so there will probably be some harmonic content, which could confuse matters considerably in many test situations. That is why capacitors C8 through C12, and coils L2 through L4 are included in the circuit. They form a multi-stage passive low-pass filter network to strip off, or at least significantly reduce, any harmonic content in the output signal. Resistors R6 through R8 form a simple attenuator/output impedance matcher, so the nominal output impedance of this RF signal generator project is 50 ohms, which is standard for much practical radio work.

Be careful if you use this RF signal generator to inject test signals into low-power stages of some radio receivers and other equipment. The signal strength may be too high. At best, this will cause severe distortion and incorrect test results. In some cases, it could damage certain semiconductor components. Use an external step-attenuator with any such project, if there is any doubt.

A metal box is recommended for enclosing any RF circuit, to provide the necessary shielding. Do not make ground connections to the case itself, especially if ac power is used to operate the project.

Keep all lead lengths as short as possible. Use a suitable shielded terminal to connect the output cable to the signal generator. Use only shielded cable, and keep it as short as possible. If using coaxial cable, remember this project is designed to have a 50-ohm output impedance. Select a coaxial cable with a compatible impedance. For some applications, you might need to use an impedance-matching transformer between the output of this RF signal generator circuit, and the input of the load circuit or device.

Project #13—Modulated RF signal injector

The previous project generates relatively pure, unmodulated RF sine waves. For some test procedures, we might need a modulated signal with a lot of harmonics. The schematic diagram for this project is shown in Fig. 7-10. A suitable parts list for

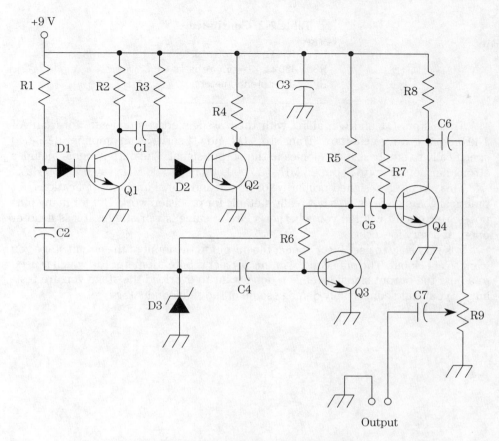

7-10 Project #12—Modulated RF signal injector.

this circuit is given in Table 7-2. Because the output signal from this circuit is so rich in harmonics, it doesn't need to be tuned. Some of the broadband signal should at least partially hit almost any RF circuit.

**Table 7-2. Suggested
parts list for Project #13—
Modulated RF signal injector of Fig. 7-10**

Q1–Q4	npn transistor (2N3904, or similar)
D1, D2	Diode (1N4148, 1N914, or similar)
D3	6.2-V zener diode 400 MW
C1–C7	0.01-μF capacitor
R1, R3	47-kΩ, ¼-W, 5% resistor
R2, R4	2.2-kΩ, ¼-W, 5% resistor
R5	1.2-kΩ, ¼-W, 5% resistor
R6	22-kΩ, ¼-W, 5% resistor

Table 7-2. Continued

R7   10-kΩ, ¼-W, 5% resistor
R8   680-Ω, ¼-W, 5% resistor
R9 2.5-kΩ potentiometer

Transistors Q1 and Q2, along with their associated components, form an AF 1-kHz square-wave generator. Transistors Q3 and Q4 serve as an amplifier designed specifically to emphasize harmonics in the output signal. The output signal includes strong harmonics up to about 25 MHz (25,000,000 Hz) to 30 MHz (30,000,000 Hz).

This circuit is designed primarily for broadband signal injector applications, for practical go/no go tests. It is not really suitable for precision work. But for many purposes, this circuit will tell you whether a circuit stage is working, or if it is dead or open, at a very low cost.

It is very easy to use. Just connect the output to the input of the circuit stage you want to check out. The only control in this project is potentiometer R9, which is used to adjust the output signal level. You don't want to overload the stage you are testing, or you could conceivably damage some of the semiconductors.

8
CHAPTER

Digital signal generation

Add enough positive feedback to almost any electronic circuit, and it will break out into oscillations. That is as true of digital gate circuits as it is for analog amplifier circuits.

A digital circuit, by definition, has just two recognizable signal levels. At any given instant, the signal is either LOW (near ground potential) or HIGH (near the supply voltage). This sounds a lot like a multivibrator as explained back in chapter 3, doesn't it? Therefore, it is natural to expect the output signal from a digital oscillator circuit to be in the form of a rectangle wave, and that is almost always the case. (Some very specialized circuit designs can force a digital circuit to generate or, more commonly, simulate a non-rectangle analog waveform, but this is, by far, the exception to the rule. Normally, unless special tricks are used, digital gates only work with rectangle waves.

In digital work, rectangle-wave generators are typically called *pulse generators* or *clocks*. Even if it is a true 1:2 duty cycle square wave, the digital technician will usually describe it as a pulse wave. In most digital applications, the actual duty cycle (and certainly the harmonic composition of the signal) doesn't matter too much, and a narrow pulse will be sufficient, so the square wave might as well be a pulse wave—it functions the same as a pulse wave.

Digital rectangle-wave generators are often referred to as *clocks* because they are so often used to synchronize the timing of events throughout a digital system of circuits. The clock signals things to happen in the right time sequence. To a circuit designer, a digital clock is not something that tells you it is time for lunch—it is just a rectangle-wave generator circuit.

Most practical digital clock circuits have a fairly narrow frequency range, and usually operate best well above the audible range. Typical digital clock frequencies are generally at least 1 MHz. Some might argue that because of the high frequencies normally encountered in such circuits, they should be included in chapter 7 (RF signal generators), rather than here. However, most digital clock circuits have special requirements and features of their own. Often, they have more in common with audio frequency signal generators than with true RF signal generators. Many of the specific circuits in this section can be set up to generate signal frequencies in the audible range.

Simple digital clock circuits

Many digital clock circuits are available to select from. Most are quite simple. Almost any digital gates can be caused to oscillate in the correct combination. In most practical digital clock circuits, gates with inverted outputs (inverters, NAND gates, or NOR gates) are often used. Even when NAND gates or NOR gates are used in such applications, they have their inputs shorted together, so they function as if they were just plain inverters, as illustrated in Fig. 8-1.

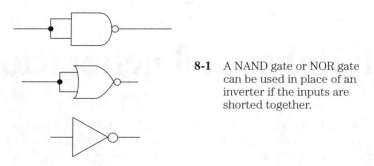

8-1 A NAND gate or NOR gate can be used in place of an inverter if the inputs are shorted together.

A typical digital clock circuit built around three inverter stages (one half a hex inverter IC) is shown in Fig. 8-2. This circuit is well-suited to use with TTL gates, such as the 7405 hex inverter chip. Because the 740-5 contains six inverter sections, two of these digital clock circuits can be constructed around a single IC. This circuit can also be built around any of the TTL subfamilies, such as the 74L05, 74H05, 74LS05, or whatever.

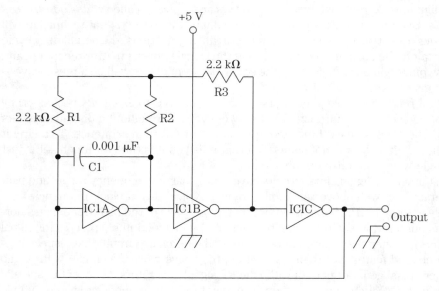

8-2 A typical digital clock circuit built around three inverter stages.

You should, however, note the fact that the 7405 hex inverter chip is used in this circuit, instead of the more commonly available 7404. This is not an arbitrary choice. The 7405's inverter stages, unlike the 7404's, have open collector outputs. This circuit might not function properly or reliably if a 7404 or similar device is used.

The output signal frequency of this clock circuit is approximately 1 MHz (1,000,000 Hz) to 10 MHz (10,000,000 Hz), depending on the value of the resistor R2. For reliable operation, this resistor should have a value in the 1-kΩ (1000 ohms) to 4-kΩ (4000 ohms) range. Resistor values outside these limits are not advisable in this circuit.

You can change the nominal frequency range of the circuit somewhat by changing the value of the capacitor. In some cases, you might find that this means you will also need to change the values of the two 2.2-kΩ (2200 ohms) resistors shown here. Unfortunately, no simple equations can be used with this circuit. Your best bet is to experiment with this circuit on a solderless breadboard and pragmatically determine the effects of various combinations of component values. Monitoring the output of the circuit with an oscilloscope is very helpful during such experimentation.

Another inverter-based digital clock circuit is shown in Fig. 8-3. This one is best suited to CMOS circuits. Only two inverter stages are required in this circuit. One third of a CD4049 hex inverter IC is an appropriate choice. Or your could use half a CD4011 quad NAND gate, or half a CD4001 quad NOR gate, with the two inputs of each gate shorted together. The left-over sections on the IC you use for this circuit can be used in other circuitry sections of the same systems, or they can be left unused. If any sections of a CMOS IC are left idle, it is very strongly recommended that all unused inputs be grounded. If the inputs are left floating, the unused section might cause instability in the other, active sections on the same chip. It wouldn't be a bad idea to ground the outputs of the unused sections as well. In some cases, a pull-down resistor is advisable between the output and ground. This will prevent problems if the output is trying to go HIGH because of its LOW (grounded) inputs.

As you can see, this circuit is extremely simple. Aside from the two inverters themselves (part of a single IC), only two external components are required—one resistor and one capacitor. These two discrete components determine the output clock fre-

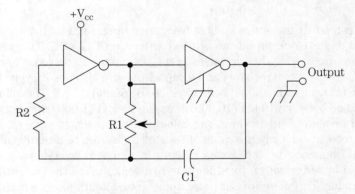

8-3 An alternate inverter based digital clock circuit.

quency. By using a potentiometer (as shown here) instead of a fixed resistor, the output frequency can be manually variable. Although not shown here, a small fixed resistor connected in series with the potentiometer is a good idea. It prevents the resistance from being set too close to zero, which could be problematic in a functional circuit.

The approximate formula for the output frequency from this circuit is:

$$F = \frac{1}{(1.4RC)}$$

where F is the frequency in Hertz, C is the capacitance in farads, and R is the resistance in ohms. For reliable operation, the capacitor's value should be kept in the 0.01 μF to 10 μF range. Increasing the value of either passive component will lower the output frequency generated by the circuit.

We'll try two quick examples to demonstrate the operational range of this digital clock circuit. First, we will try very small component values, to give us the approximate maximum frequency. Let's say R has a value of 1.2-kΩ (1200 ohms), and C is 0.01 μF (0.00000001 farad). In this case, the output frequency works out to a value of about:

$$F = \frac{1}{(1.4 \times 1200 \times 0.00000001)}$$

$$= \frac{1}{0.0000168}$$

$$= 59{,}524 \text{ Hz}$$

or about 60 kHz.

At the opposite extreme, let's say R is set for a resistance of 500-kΩ (500,000 ohms), and C is a 10-μF (0.00001 farad) capacitor. This time, the frequency (approximately its lowest possible value) will be about:

$$F = \frac{1}{(1.4 \times 500000 \times 0.00001)}$$

$$= \frac{1}{7}$$

$$= 0.14 \text{ Hz}$$

This circuit can cover the entire audible frequency range, and then some.

Our next digital clock circuit can be used with either TTL or CMOS gates. Two inverter sections are used in this circuit, as in the last circuit discussed. Three external passive components—two resistors and a capacitor—are also used. Typically, the capacitor value in this circuit should be between 0.01 μF and 0.1 μF. The value of resistor R2 should be between 10 kΩ (10,000 ohms) and 1 MΩ (1,000,000 ohms) and R1's value should be about 5 to 10 times the value of R2. The exact value of resistor R1 usually isn't too critical. Only the value of resistor R2 is used to directly calculate the output signal frequency. In some cases, resistor R2 can be replaced by a potentiometer connected in series with a fixed limiting resistor to make the output frequency manually variable. The values of this potentiometer and limiting series resistor should be selected so that their series combination will always be between about one-tenth and one-fifth of the value of resistor R1. You can usually get away with a little bit of "fudging" here, but not too much, or the circuit will not function reliably.

The approximate output clock frequency from this circuit can be calculated with this formula:

$$F = \frac{R_2 C}{2.2}$$

This equation is just an approximation. Don't expect it to give exact results, even ignoring the effects of component tolerances. Unlike most such equations, the value of C should be in microfarads, not farads. R_2 should be in ohms, and F will be in Hertz.

Let's try a couple examples. First, let's use the minimum recommended component values—a 10-kΩ (10,000 ohms) resistor for R_2, and a 0.01-μF capacitor for C. We can use a 68-kΩ resistor for R_1. Anything from a 50-kΩ to 100-kΩ resistor would do. The approximate output frequency in this case will be about:

$$F = \frac{(10000 \times 0.01)}{2.2}$$

$$= \frac{100}{2.2}$$

$$= 45 \text{ Hz}$$

At the opposite extreme, we will use a 1-MΩ (1,000,000 ohms) resistor for R2, and a 0.1-μF capacitor for C. A 10-MΩ resistor would probably be the best choice for R1, because resistors above 5 MΩ but less than 10 MΩ are very difficult to find, even though they would work electrically. This time the output frequency works out to approximately:

$$F = \frac{(1000000 \times 0.1)}{2.2}$$

$$= \frac{100000}{2.2}$$

$$= 45,455 \text{ Hz}$$

or about 45 kHz.

This circuit can't generate the lowest audible frequencies (the nominal audible frequency range goes down to about 20 Hz), but it comes close. Its upper limit is well above the highest audible frequency limits.

We can use NAND gates with the inputs shorted together in place of the two inverters in this last circuit, of course. This is shown in Fig. 3-4. This is not a trivial repetition of the general principle we've mentioned several times already in this section. The use of NAND gates in this circuit permit us to make an interesting modification in the basic circuit, as illustrated in Fig. 8-5.

Unlike most oscillator circuits, this one has an external input. One of the NAND gate inputs can be externally controlled by any appropriate digital signal. This will automatically turn the clock circuit on and off, as desired. A logic 0 (LOW) input signal will inhibit the signal generator's operation. There will be no output signal from this circuit. A logic 1 (HIGH) input signal will enable the digital clock, and it will generate a rectangle-wave clock signal at its output in the usual way.

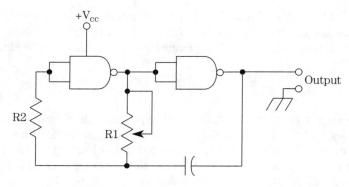

8-4 A pair of NAND gates with the inputs shorted together can take the place of the two inverters in the circuit of Fig. 8-3.

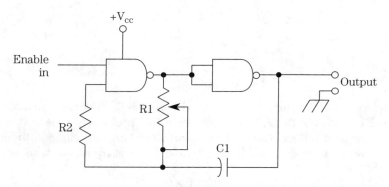

8-5 This variation of the clock circuit of Fig. 8-4 can be gated by an external digital signal.

In the analog oscillator circuits of chapter 2, we learned that the most precise oscillation frequencies can be achieved if a crystal is used as the primary frequency-determining component. The same thing is also true in digital oscillator circuits, except that the output signal will be in the form of a rectangle wave, instead of a sine wave. Figure 8-6 shows a simple digital crystal oscillator circuit, using NOR gates. Notice that a NOR gate must be used for the upper gate device in the schematic diagram. Two inputs are required, and a NAND gate won't work properly in this application. The second gate section can be a regular inverter, but we are already using one NOR gate in this circuit, and we might as well use a second NOR gate (with its inputs shorted together) here.

This is a very precise circuit, thanks to the use of the crystal. Of course, the crystal itself is the primary factor in determining the circuit's output frequency. For more general information on crystals and their frequencies, refer back to the section on (analog) crystal oscillators in chapter 2. Varying the value of capacitor C (typically from 4 pF to 40 pF) permits some minor fine-tuning of the output frequency for precision applications.

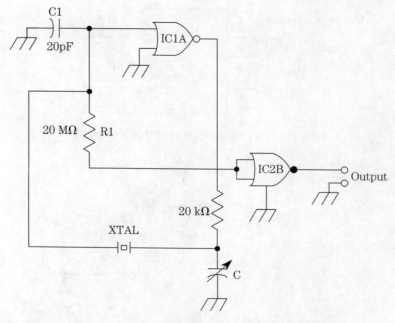

8-6 A simple digital crystal oscillator circuit, using NOR gates.

The component values indicated in the schematic diagram were selected for peak performance for output frequencies near 1 MHz (1,000,000 Hz). For other output frequencies, you might want to experiment with other component values on a solderless breadboard. Monitor the output signal with an oscilloscope and select the component values to give the cleanest possible output signals.

As with most analog crystal oscillators, this circuit isn't too well-suited for output frequencies within the audible range.

Dedicated clock ICs

Dedicated digital clock ICs are also available. They usually offer a number of special features, and are significantly more sophisticated than any of the simple gate circuits we have worked with here. For example, Intersil makes a general-purpose CMOS timer chip called the ICM7209. The pin-out diagram for this device is shown in Fig. 8-7. A typical clock circuit built around the ICM7209 is shown in Fig. 8-8. It requires a minimum of three external components—two capacitors and a crystal. This circuit is best suited for crystal frequencies from 10 kHz (10,000 Hz) up to 10 MHz (10,000,000 Hz). It does cover the upper half of the audible frequency range, but it is used mostly for frequencies much higher than the audible limits.

This particular clock circuit's frequency is not manually tunable. To change the output frequency, it is necessary to replace the crystal. In many applications, multiple crystals that are individually switch-selectable might provide a reasonable degree of control.

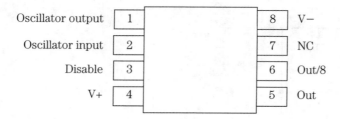

8-7 The ICM7209 is a dedicated digital clock IC.

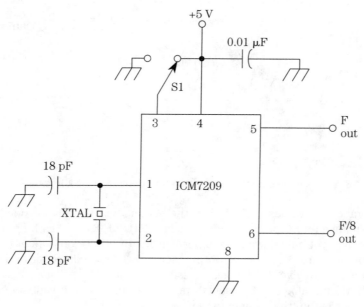

8-8 A typical clock circuit built around the ICM7209.

Variable-duty cycle rectangle-wave generators

Most digital clock circuits produce square waves. In most digital applications, the duty cycle isn't of any particular importance. Because a square wave is usually the easiest and least expensive to generate, there is no real reason to bother about the duty cycle.

But there will always be those occasional oddball applications that might require some duty cycle other than the simple 1:2 of a square wave. Fortunately, it is not too difficult to modify some digital oscillator circuits to change the duty cycle of the generated rectangle wave.

When you read about inverters in the circuits described in this section, remember, a digital inverter can be created by shorting together the inputs of a NAND gate or a NOR gate. Electrically, there is no difference in such a change. This flexibility permits you to use almost any spare digital gates you might have free from other circuitry in a larger system.

The basic digital square wave oscillator circuit we will be working with is shown in Fig. 8-9. The output signal frequency from this circuit is determined by the values of resistor R1 and capacitor C1. This circuit is suitable for generating signal frequencies ranging from about 1 Hz to over 1 MHz (1,000,000 Hz, or 1,000 kHz). Quite an impressive range.

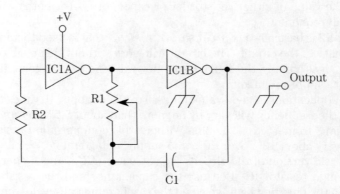

8-9 This basic digital square-wave oscillator circuit will be the starting point for the circuits throughout this section.

For reliable operation, resistor R1 must have a value of at least 47 kΩ (47,000 ohms), but no more than 22 MΩ (22,000,000 ohms). The capacitor value can range from a few dozen picofarads (pF) up to several microfarads (μF). As a reminder, a microfarad is one millionth of a farad, and a picofarad is one millionth of a microfarad. The capacitor must be a non-polarized capacitor. A polarized capacitor, such as an electrolytic capacitor or tantalum capacitor, will not work in this circuit.

The approximate output signal frequency for this circuit can be found using this simple formula:

$$F = \frac{1}{(1.4 R_1 C_1)}$$

As an example, let's say resistor R1 has a value of 22 kΩ (22,000 ohms), and capacitor C1 has a value of 0.0047 μF (0.0000000047 farad). In this case, the output signal frequency will be about:

$$F = \frac{1}{(1.4 \times 22000 \times 0.0000000047)}$$

$$= \frac{1}{0.0001447}$$

$$= 6,911 \text{ Hz}$$

Now, let's keep C1 at 0.0047 μF, but increase R1 to 680 kΩ (680,000 ohms). This will change the circuit's output signal frequency to about:

$$F = \frac{1}{(1.4 \times 680000 \times 0.0000000047)}$$

$$= \frac{1}{0.0044744}$$

$$= 223 \text{ Hz}$$

Increasing the value of either resistor R1 or capacitor C1 (or both) will reduce the output signal frequency.

Resistor R2 is basically a *compensation resistor* that is used to improve the frequency stability of the circuit. The circuit will work without this resistor, but a little less reliably. Most notably, the signal frequency will tend to change with fluctuations of the circuit's supply voltage.

In most applications, a good value for resistor R2 is about 10 times that of R1. In such a case, the frequency will vary by no more than about 0.5% for any supply voltage fluctuations from +5 to +15 volts. Without this compensation resistor, the frequency will vary about 12% over the same supply voltage range.

In this basic version of the circuit, the capacitor (C1) is both charged and discharged through resistor R1. Because the capacitance and the resistance are the same for both the charging half-cycle and the discharging half-cycle, the two half-cycles must logically have identical time periods. Therefore, the duty cycle will be 1:2, and the output signal will be a square wave.

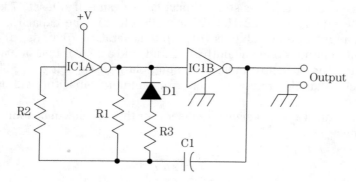

8-10   A very simple non-square rectangle-wave signal generator circuit.

This suggests the obvious approach to varying the duty cycle—we need to use different resistance for the charging and discharging half-cycles. A very simple non-square rectangle wave signal generator circuit is shown in Fig. 8-10.

When the capacitor is charging, the diode is forward-biased, and the resistance seen by the capacitor (determining the timing period) will be equal to the parallel combination of resistors R1 and R3. For convenience, we can reasonably ignore the small internal resistance of the forward-biased diode itself. The formula for the two resistances in parallel is, as always:

$$R_t = \frac{(R_1 \times R_3)}{(R_1 + R_3)}$$

The HIGH time period will be about:

$$T_h = \frac{0.7C1 \ (R_1 \times R_3)}{(R_1 + R_3)}$$

On the LOW half cycles, when the capacitor is discharged and re-charged with the opposite polarity, the diode is reverse-biased. It's very high resistance can effectively be considered an open circuit, so resistor R3 is not a functional part of the circuit. Therefore, the time period for this half-cycle is equal to approximately:

$$T_l = 0.7C_1R_1$$

The total cycle time is simply the sum of the LOW and HIGH timing periods:

$$T_t = T_h + T_l$$

$$= \frac{0.7C_1(R_1 \times R_3)}{(R_1 + R_3) + 0.7C_1R_1}$$

$$= \frac{1.4C_1R_1(R_1 \times R_3)}{(R_1 + R_3)}$$

The signal frequency is always equal to the reciprocal of the total cycle time:

$$F = \frac{1}{T_t}$$

$$= \frac{1}{[1.4C_1R_1(R_1 \times R_3) \ / \ (R_1 + R_3)]}$$

For example, let's assume we are using the following component values in the circuit:

$C1 = 0.001 \ \mu F$ (0.000000001 farad)
$R1 = 100 \ k\Omega$ (100,000 ohms)
$R2 = 1 \ M\Omega$ (1,000,000 ohms)
$R3 = 27 \ k\Omega$ (27,000 ohms)

Plugging these component values into the equations, we find:

$$T_h = \frac{0.7C_1(R_1 \times R_3)}{(R_1 + R_3)}$$

$$= \frac{0.7 \times 0.000000001 \times (100000 \times 27000)}{(100000 + 27000)}$$

$$= \frac{0.0000000007 \times 2700000000}{127000}$$

$$= 0.0000000007 \times 21260$$

$$= 0.000015 \ \text{second}$$

$$= 0.015 \ \text{mS}$$

$$= 15 \ \mu S$$

$$T_l = 0.7C_1R_1$$
$$= 0.7 \times 0.000000001 \times 100000$$
$$= 0.00007 \text{ second}$$
$$= 0.07 \text{ mS}$$
$$= 70 \text{ } \mu\text{S}$$

$$T_t = T_h + T_l$$
$$= 0.000015 + 0.00007$$
$$= 0.000085 \text{ seconds}$$
$$= 0.085 \text{ mS}$$
$$= 85 \text{ } \mu\text{S}$$

$$F = \frac{1}{T_t}$$
$$= \frac{1}{0.000085}$$
$$= 11,765 \text{ Hz}$$

The duty cycle, of course, is defined by the relative charge and discharge resistances.

Notice that in this circuit, the HIGH half-cycle is always shorter than the LOW half-cycle.

A better, variable-duty cycle digital rectangle-wave oscillator circuit is shown in Fig. 8-11. The charge and discharge resistances can be independently set in this circuit. At any given instant, one diode is forward-biased and the other is reverse-biased. On one half-cycle, the timing resistance is the series combination of R1 and R3.

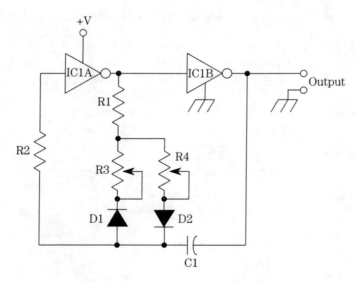

8-11 This is a better, variable duty-cycle digital rectangle-wave oscillator circuit.

On the other half-cycle, the timing resistance is the series combination of R1 and R4. Remember, resistances in series are simply added together.

The problem with both of these rectangle-wave signal generators is that changing the duty cycle also affects the signal frequency.

The digital rectangle-wave signal generator circuit shown in Fig. 8-12. During one half-cycle, diode D1 is forward-biased, and diode D2 is reverse-biased. On the other half-cycle, diode D1 is reverse-biased, and diode D2 is forward-biased. On each half-cycle, the capacitor "sees" a different resistance, but the total circuit resistance remains constant. Adjusting potentiometer R3 changes the duty cycle, but the total frequency-determining resistance in the circuit isn't changed, so the signal frequency is not affected. Actually, there might be some slight frequency change, but it will be a lot less than in the earlier circuits.

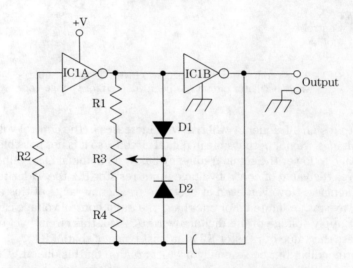

8-12 This digital rectangle-wave signal generator circuit permits the duty cycle to be changed without affecting the signal frequency.

Digital VCOs

Earlier, we saw that many analog oscillator and signal generator circuits can be adapted for voltage control. A VCO makes many new applications possible, including automated functions of various types.

It's not too common, but a digital VCO is not only possible, but surprisingly easy to achieve. A variation on the basic two-inverter digital square wave oscillator circuit adapted for voltage control is shown in Fig. 8-13. The base frequency of this circuit is determined in the usual way:

$$F = \frac{1}{(1.4C_1R_1)}$$

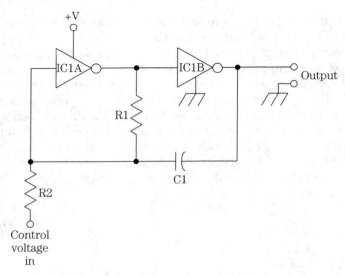

8-13   This variation on the basic two-inverter digital square-wave oscillator circuit has been adapted for voltage-control.

The output signal frequency will rise with increases in the control voltage input. Negative voltages are not permissible in digital circuits, so it is not possible to use the control voltage to lower the signal frequency below its nominal base value.

The lower the value of control voltage input resistor R2, the wider the range of output frequencies. However, even at best, the frequency range of this digital VCO circuit is always rather limited. For one thing, the input control voltage cannot safely exceed the supply voltage of the digital inverters. Also, this circuit will not operate reliably unless the value of resistor R2 is at least twice of that of R1.

A rather peculiar, but occasionally useful variation on this digital VCO circuit is shown in Fig. 8-14. Again, increasing the control voltage input will increase the VCO's output frequency. However, if the control voltage drops below a specific pre-set level, the oscillator's output will be cut off altogether. This pre-set cut-off point is set by adjusting potentiometer R2.

The CD4046 PLL

A very interesting digital device that is useful for many signal-generation applications is the CD4046 digital *PLL*, or *phase-locked loop*. This device is underused by most electronics hobbyists, although it is widely used in many commercially manufactured circuits.

Phase-locked loops are used in many analog applications too, including, but not limited to signal generation. Unfortunately, an extensive study of these devices is not within the scope of this book, but I will devote some space here to the CD4046, as it applies to signal-generator applications.

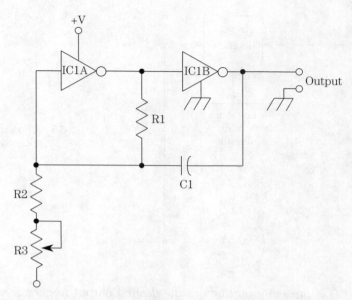

8-14 This variation on the digital VCO circuit of Fig. 8-13 will cut itself off whenever the control voltage drops below a specific pre-set level.

Probably the chief reason the CD4046 is so underused in hobbyist electronics is that so many people think that PLLs are very mysterious and difficult to understand. This idea isn't really accurate. Sure, PLL circuits can get very complicated and sophisticated, and it does take a lot of heavy math to use a PLL to its full capabilities. But the basic principles are simple enough, and it is easy enough to use PLLs in IC form in many relatively simple-to-moderate applications—even if the most high-tech applications that are theoretically possible might be a bit out of the grasp of the average electronics hobbyist.

A phase-locked loop is essentially a pseudo-servo-system with a controlling feedback loop. You might call it a sort of self-correcting VCO. A basic PLL is made up of three functional stages, as illustrated in Fig. 8-15. The output frequency of the PLL locks onto and follows (or tracks) an input reference signal. The output frequency is not necessarily equal to the frequency of the input reference signal, but the output frequency is always an integral multiple of the input frequency.

The VCO generates a signal with some specific basic frequency, determined by component values in the circuit. But this base frequency does not have to be precisely set. That is the whole point of the PLL. The VCO's output signal frequency might be higher than it should be, lower than it should be, or exactly on the desired frequency. The circuit designer or user doesn't have to worry much about it. The PLL automatically corrects its own output signal frequency. Some of the VCO's output frequency is fed back to the PLL's phase detector stage, where it is compared with the input reference signal.

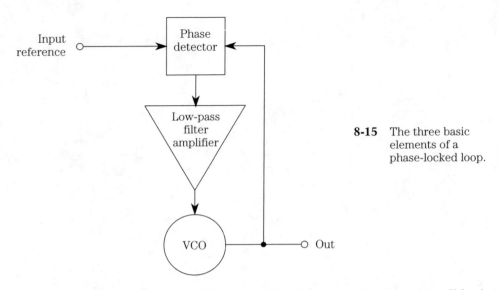

8-15 The three basic elements of a phase-locked loop.

If the VCO is presently putting out the desired output frequency, it will be in phase with the input signal, and the phase detector won't put out an error voltage. Or, if you prefer, you can say that the error voltage is zero. The VCO will continue to generate the output signal at the same frequency. Nothing will be changed.

If the VCO's output signal frequency is an exact whole number multiple of the input reference signal frequency, they can still be properly in phase with one another, as illustrated in Fig. 8-16.

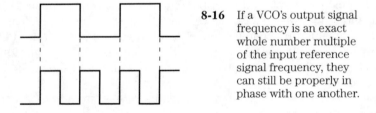

8-16 If a VCO's output signal frequency is an exact whole number multiple of the input reference signal frequency, they can still be properly in phase with one another.

On the other hand, let's suppose that the output signal from the VCO drops to too low a frequency. It will necessarily be out-of-phase with the input reference signal, because their frequencies are not exact multiples of one another. The phase detector will put out an error voltage that is proportionate to the degree of detected phase shift. This error signal is conditioned and smoothed out by the filter/amplifier stage, then it is fed into the VCO's input, forcing it to raise the frequency of its output signal. This process will continue until the feedback signal from the VCO's output properly matches up to the reference input signal at the phase detector.

Pretty much the same thing happens if the VCO's output frequency goes too high for any reason. The only difference is that in this case, the generated error voltage will have the opposite polarity, so it will shift the VCO's frequency in the opposite direction—that is, lower.

Typical applications for PLLs include:
- Frequency demodulation
- Frequency modulation
- Frequency synthesis
- FSK demodulation
- Signal generation
- Tone decoding
- Voltage-to-frequency conversion (that is, VCOs)

The pin-out diagram for the CD4046 digital PLL IC is shown in Fig. 8-17. Figure 8-18 is a rough block diagram of the CD4046's internal circuitry. Notice that this chip contains two-phase comparator stages. Phase comparator A is actually just a simple exclusive-OR (X-OR) gate. It provides good noise immunity, but at a price. This comparator will sometimes incorrectly lock onto input signals with frequencies close to one of the harmonics of the VCO's output signal. Another disadvantage with this phase comparator is that it only works properly if the input reference signal is a good, clean square wave, with a true 50% duty cycle, or very close to it.

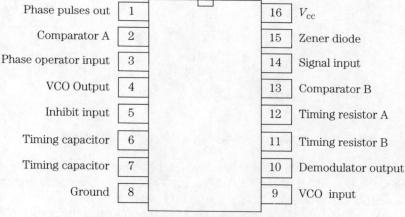

Phase pulses out	1	16	V_{cc}
Comparator A	2	15	Zener diode
Phase operator input	3	14	Signal input
VCO Output	4	13	Comparator B
Inhibit input	5	12	Timing resistor A
Timing capacitor	6	11	Timing resistor B
Timing capacitor	7	10	Demodulator output
Ground	8	9	VCO input

8-17   The CD4046 is a digital PLL in IC form.

Because of these problems, an alternate phase comparator is included within the CD4046 chip. Phase comparator B is made up of somewhat more complex circuitry, and is less prone to harmonic lock-up, but its noise immunity isn't quite as good as for phase comparator A. Phase comparator B has a wider tracking range than phase comparator A. Its range is better than 1000:1. In addition, this phase comparator can function with input pulses with almost any duty cycle, whereas phase comparator A is limited to 50% duty cycle square waves. Both phase comparators do recognize only rectangle wave input signals, as you'd expect with any digital device.

Obviously, selecting which phase comparator to use in a given circuit calls for a compromise. The best choice will depend on the particular requirements of the intended application. By making both phase comparators available, the circuit de-

8-18 A rough block diagram of the CD4046's internal circuitry.

signer is allowed to make the selection himself. If he wants to use phase comparator A, he takes the output off of pin #2, or if he prefers to use phase comparator B, he takes the output off of pin #13. That is certainly simple enough.

An inhibit input (pin #5) is provided for both the VCO and the source follower. A logic 0 (LOW) on this pin enables these stages, while a logic 1 (HIGH) turns them off, reducing power consumption. The CD4046 PLL also includes an on-chip 5.2 volt zener diode (between pin #15 and ground—pin #8) for simple voltage-regulation applications. Its use is optional.

The supply voltage for the CD4046 should be between +3 and +18 volts. A regulated supply voltage isn't absolutely essential, but it is a good idea. Current drain by this chip is quite low. The exact value will depend on the VCO frequency and the percentage of time it is enabled by pin #5. Typically, power consumption will be about $\frac{1}{160}$ of that required for a standard linear PLL IC, such as the 565. The CD4046's low-power consumption makes it practical for use in battery-operated applications.

In signal-generation applications, the VCO is the most important and interesting stage in the CD4046. It can generate output frequencies up to 1 MHz (1,000,000 Hz) and beyond. The output signal is a very symmetrical square wave. The duty cycle of the output signal is always 1:2. The voltage-to-frequency linearity of this VCO is excellent. A typical linearity rating for the VCO is 1%. The control voltage can drive the VCO through a 1:1,000,000 range.

The base frequency of the CD4046's VCO is determined by a capacitor connected between pins #6 and #7, and a resistor connected between pin #11 and ground. The minimum recommended value for this timing capacitor is 50 pF. The timing resistor's minimum recommended value is 10 kΩ (10,000 ohms).

The output signal frequency can be shifted from its nominal value by an external control voltage, of course. The control voltage input is applied to pin #9. In PLL applications, the control voltage will be taken from one of the on-chip phase comparators. For signal generation VCO applications, the phase comparators are usually left unused and ignored.

A simple square-wave signal generator circuit built around the CD4046 PLL IC is shown in Fig. 8-19. This is probably one of the simplest circuits that uses this chip. Only three external components are required—a capacitor, a fixed resistor, and a potentiometer. The potentiometer serves as a simple voltage divider. Adjusting this potentiometer changes the control voltage seen by the CD4046's VCO, and thus its output signal frequency. Using the component values shown in the schematic diagram, the output frequency from this circuit can range from a little over zero (dc) up to about 5 kHz (5,000 Hz). For best results, the supply voltage used to power this circuit should be well-regulated, or the signal frequency might be inclined to drift unacceptably.

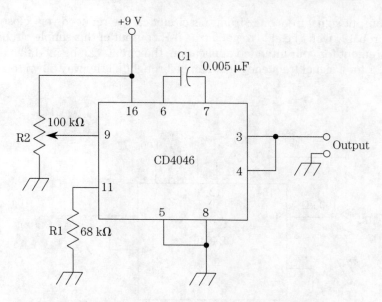

8-19 This simple square-wave signal generator circuit is built around the CD4046 PLL IC.

To adapt this square-wave signal generator circuit for a true VCO application, you simply have to omit the voltage-divider potentiometer, and replace it with any suitable external control voltage, as shown in Fig. 8-20. The control voltage should never be permitted to exceed the supply voltage applied to the CD4046.

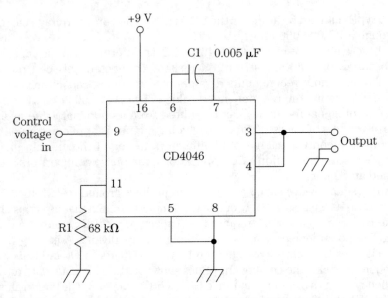

8-20   The square-wave signal generator circuit of Fig. 8-19 can be easily adapted for a true VCO application.

The output signal frequency from this circuit can be reduced very close to zero, but you probably won't be able to get a true 0-Hz output of this simple circuit. If this is a requirement for your intended application, the circuit can be modified as shown in Fig. 8-21 for a signal frequency that can be brought all the way down to zero.

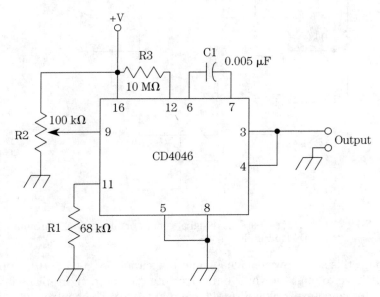

8-21   The signal frequency of this circuit can be brought all the way down to true zero.

This circuit is a little more complex than the one shown in Fig. 8-19, but it is still a remarkably simple signal generator circuit. The secret here is the resistor added between pin #12 and the supply voltage. Notice that this resistor is given a very high value. When the output signal frequency is reduced to zero, what will the output state be—HIGH or LOW? That is impossible to predict. The circuit's output will randomly settle into one state or the other.

In other applications, you might want to set a specific limit to the lowest output signal frequency the circuit can be made to put out. This can be done with the circuit illustrated in Fig. 8-22. Using the component values suggested here, the output frequency range will be limited to about 70 Hz to 5 kHz (5,000 Hz). The minimum output frequency is determined by the values of capacitor C1 and the setting of potentiometer R2, or the externally applied control voltage. The maximum output frequency is determined by C1, R1, and R2. In effect, R1 and R2 function as if they were in parallel with one another.

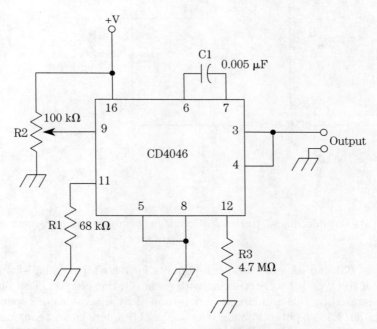

8-22   In this variation of the circuit of Fig. 8-20, there is a preset minimum output frequency.

Figure 8-23 demonstrates how the CD4046's inhibit input (pin #5) can be used for gated operation of a square wave signal-generator circuit. Because the inhibit input is buffered by a digital inverter stage, the circuit is activated (enabled) by a HIGH gate input, and disabled (inhibited) by a LOW gate input. Otherwise, this circuit works pretty much like the others we have discussed in this section.

The circuit shown in Fig. 8-24 automates the inhibit function of the CD4046 in a recurring pattern. The output is in the form of bursts of a tone, separated by equal-length pauses.

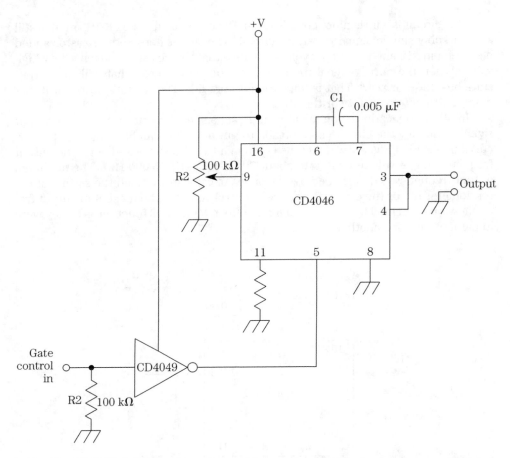

8-23 The CD4046's inhibit input can be used for gated operation of a square-wave signal generator circuit.

Basically, IC1 and its associated components form a simple square-wave signal generator, of the type we were working with earlier in this chapter. The signal frequency generated by this sub-circuit is determined by the values of capacitor C1, and resistor R1. R1 is a potentiometer, so the signal frequency can be manually varied. A fixed resistor could be substituted, of course, if that better suits your specific intended application. Because capacitor C1 is relatively large, this oscillator will generate a very low, sub-audible frequency. The output from this signal generator sub-circuit controls the inhibit/enable input (pin #2) of the CD4046 (IC2).

The CD4046 itself is wired just as a standard square-wave generator VCO. The control voltage is derived from a simple voltage divider string comprised of resistors R3 through R5. R4 is a potentiometer—and the operator of this circuit can use it to manually control the tone frequency. The output signal from pin #4 of IC2 can be fed through an audio amplifier and loudspeaker, or it could be used to drive some other load circuit or device.

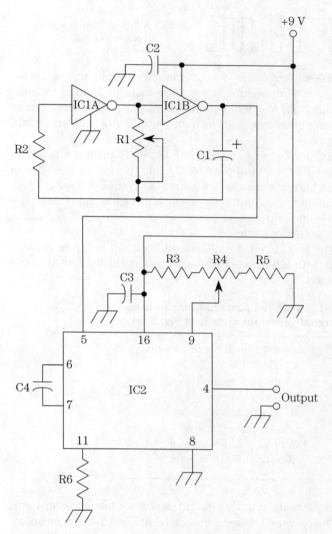

8-24 This circuit automates the inhibit function of the CD4046.

Potentiometer R4 controls the tone frequency or pitch that will be heard if the output signal is fed through a loudspeaker. Potentiometer R1 controls the timing of the tone-on/tone-off bursts. A typical output signal from this circuit is illustrated in Fig. 8-25. Notice that during the between-tone pauses, the signal level might be held HIGH or LOW. This is random, and cannot be predicted. This built-in randomized effect of the use of the CD4046's inhibit function could limit the practical use of this circuit in some control applications.

If the inhibit-control frequency (generated by IC1) is brought up into the audible range, the result will be a form of modulation of the CD4046's output signal. The human ear will no longer be able to distinguish between the separate bursts of tone, because they are too close together. The effect will be a rather raspy, continuous

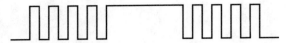

8-25 A typical output signal from the circuit of Fig. 8-24.

tone. Because both signals are square waves, a great many sidebands will be generated by this modulation process. I'm not sure just what to call this form of modulation—it isn't amplitude modulation, frequency modulation, pulse-code modulation, pulse-width modulation, or any other standard form of modulation. Still, you might find it useful for certain purposes.

Capacitors C2 and C3 are simple power supply line filters, to protect the CMOS ICs from possible noise spikes that might appear on the supply voltage line. Use of such filter capacitors is strongly recommended with any CMOS IC. For the best protection, the filter capacitor should be mounted as physically close as possible to the IC it is protecting. The exact value of this capacitor is not critical, and will not affect the circuit's operation in any noticeable way.

A typical parts list for the tone burst signal-generator circuit is given in Table 8-1. Feel free to experiment with alternate component values throughout this circuit. Nothing is terribly critical here.

Table 8-1. Suggested parts list for the tone burst signal generator circuit of Fig. 8-24

IC1	CD4049 hex inverter (see text)
IC2	CD4046 PLL
C1	10-μF, 25-V electrolytic capacitor
C2, C3	0.01-μF capacitor
C4	0.001-μF capacitor
R1, R4	250-kΩ potentiometer
R2	3.3-MΩ, ¼-W, 5% resistor
R3, R5, R6	100-kΩ, ¼-W, 5% resistor

Only two inverter sections are used in this circuit, so this is just one-third of a CD4049 hex inverter chip. The four unused sections can be utilized in other circuitry as part of a larger system, or they can be left unused. If left idle, it is a good idea to ground any unused inputs. If the inputs of the unused sections are left floating, all of the gates on the same chip could become unstable, including the ones you are actively using. As always, the actual inverters can be replaced with NAND gates or NOR gates with their inputs shorted together.

Project #14— Digital odd-waveform signal generator

It might seem that the signals that the CD4046 PLL's VCO generates would get rather boring after awhile. After all, the output signal from this chip is always a square wave. You can change the frequency, or you can turn it on or off, and that's about it. Right?

Well, yes and no.

The circuit shown in Fig. 8-26 generates a wide variety of very unusual signals. It is fascinating to feed the output from this project to an audio amplifier and loudspeaker and listen to the changing sound effects as you adjust one or more of the five

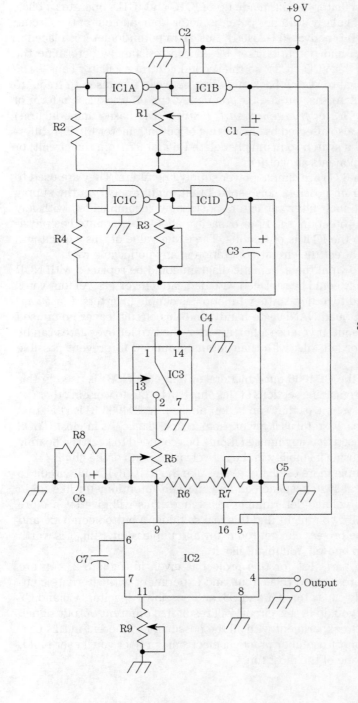

8-26 Project #13—Digital odd-waveform signal generator.

potentiometers (R1, R3, R5, R7, and R9) in this circuit. With five potentiometers, the number of possible combinations is almost mind-boggling.

One control oscillator is made up of IC1A and IC1B. The output signal from this oscillator stage controls the CD4046's (IC2) inhibit input (pin #5) as discussed above.

A second control oscillator circuit, made up of IC1C and IC1D, operates a bilateral switch (IC3), that regularly switches potentiometer R5 in and out of the circuit. When the bilateral switch is activated (closed), this extra potentiometer is placed in parallel with resistor R6 and potentiometer R7. These resistances determine the control voltage fed into the VCO (IC2), so the second control oscillator causes the circuit's final output signal to switch back and forth between two discrete frequencies. If this control oscillator has an audible frequency, the effect will be a form of modulation, known as *FSK*, or *frequency shift keying*. The base (unmodulated) frequency of the VCO is also affected by the setting of potentiometer R9. For different effects, you can add a switch to manually delete R6 and R7 from the circuit, or you can omit these components altogether.

Capacitors C2, C4 and C5 are simple power supply line filters. They are used to protect the CMOS ICs from possible noise spikes that might appear on the supply voltage line. The use of such filter capacitors is strongly recommended with any CMOS IC. For the best protection, the filter capacitor should be mounted as physically close as possible to the IC it is protecting. The exact value of this capacitor is not critical, and will not affect the circuit's operation in any noticeable way.

The NAND gates shown in the schematic diagram could be replaced with NOR gates, if you prefer, or they could be replaced with dedicated inverters. All four gates have their inputs shorted together, so they function as simple inverters. I just happened to have a CD4011 quad NAND gate handy, and this circuit leaves no unused gate sections to worry about. If you use a hex inverter, the two leftover gates can be used in other circuitry, or left idle, with their inputs grounded to prevent possible stability problems.

Only one section of the CD4066 quad bilateral-switch chip (IC3) is used in this project. Again, the three remaining sections of this chip can be put to work in other circuitry as part of a larger system, or they can be left unused. A CD4066 IC isn't all that expensive, so this is not as wasteful as it might seem at first glance. As in most CMOS circuits, it is strongly advised that any unused inputs be grounded to keep the floating sections from possibly causing instability in the actively used parts of the chip.

You could build a couple more oscillator stages like the IC1C/IC1D sub-circuit to control additional bilateral switches, each adding another potentiometer like R5 in parallel with the control voltage-determining resistance. This will give even more complex patterns. Actually, you might find this makes things a bit too complex, and the tone generated by the project can get too raspy and unpleasant. Still, it is worth experimenting with. Who knows? You might like it.

A complete suitable parts list for this project is given in Table 8-2. You are strongly encouraged to breadboard this circuit and experiment with alternate component values—especially capacitors C1 and C3, and possibly capacitors C6 and C7 as well. Almost anything you do in this circuit will result in a different output signal, and it can be fascinating to experiment with all the possibilities inherent in this project. It might not be the most useful or practical electronic project you'll ever build, but it could certainly be one of the most fun.

**Table 8-2. Suggested parts list for Project #14
—Digital odd-waveform signal generator of Fig. 8-26**

IC1	CD4011 quad NAND gate (see text)
IC2	CD4046 PLL
IC3	CD4066 quad bilateral switch (one section only)
C1, C6	1-µF, 25-V electrolytic capacitor
C2, C4, C5	0.01-µF capacitor
C3	4.7-µF, 25-V electrolytic capacitor
C7	0.001-µF capacitor
R1, R3	100-kΩ potentiometer
R2, R4	1-MΩ, ¼-W, 5% resistor
R5, R7	500-kΩ potentiometer
R6	220-kΩ, ¼-W, 5% resistor
R8	470-kΩ, ¼-W, 5% resistor
R9	250-kΩ potentiometer

Frequency multiplication and division

In some applications, we might want a signal generator to do double duty. We might need two related frequencies. The obvious solution is to simply build two signal-generator circuits, each designed for its own specific frequency. Besides being rather inelegant, and requiring a larger circuit cost, this approach is inadequate in any applications where exact timing is critical. It is difficult, at best, to fine-tune the two separate signal generators precisely together, and their output signals will almost certainly be out-of-phase with one another.

A more elegant and effective solution is to somehow derive the second signal frequency directly from the first. It can require some tricky circuitry to do such things in most analog systems, but when working with digital signals, it's a snap.

If we want to increase the signal frequency, we use a frequency multiplier circuit, or, if we want to decrease the signal frequency, we use a frequency divider circuit. That's obvious enough. In practical circuits, we are usually limited to whole number multipliers and dividers. That is, both the input and output signal frequencies will be harmonically related.

We will look at frequency multiplication first. It is a little trickier than frequency division, especially for odd integer values. It isn't too much of a problem, however, to multiply a signal frequency by a power of two. A simple "times-two" frequency multiplier circuit is shown in Fig. 8-27. As you can see, there isn't much to this circuit. It consists of just four X-OR gates, and a 2.2-kΩ (2200 ohms) resistor. All four gates can be in a single CD4070 quad X-OR IC, so the total component count for this circuit is just two.

You might want to add a third component—a line filter capacitor between the IC's supply voltage pin and ground. This capacitor will help protect the CMOS circuitry inside the IC from possible noise spikes in the supply voltage line. It should be mounted as physically close as possible to the body of the chip. The exact value of this capacitor is not at all critical and will not affect the operation of the circuit in any

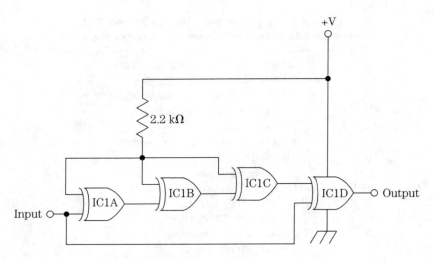

8-27 A simple "times-two" frequency multiplier circuit.

noticeable way. Typical values range somewhere between 0.01 μF to 0.1 μF. Just use whatever capacitor in this range (or close to it) that you happen to have handy.

An X-OR gate is also known as an *Exclusive-OR gate*. As the name suggests, it is a variation on the basic OR gate. A regular OR gate has a HIGH (1) output whenever any of its inputs are HIGH. The output is LOW (0) if, and only if, all inputs are LOW. The truth table for a standard two-input OR gate is as follows;

Inputs		Output
A	B	
0	0	0
0	1	1
1	0	1
1	1	1

An X-OR gate normally has just two inputs. The output of this gate is HIGH if either input A or input B is HIGH, but not both. This is the exclusive part. As the X-OR gate's truth table suggests, it is a one-bit digital comparator, also known as a *difference detector*:

Inputs		Output
A	B	
0	0	0
0	1	1
1	0	1
1	1	0

The output of the X-OR gate is HIGH when its inputs are at opposite states. If the two inputs are identical, the output of the gate will be LOW.

In our frequency-multiplier circuit, "input A" to the first three gate stages is permanently held HIGH through pull-up resistor R1. That means these inputs can never be in the LOW state in this circuit.

If you try to trace through the logic states in this circuit, you will find that the inputs to the final gate stage (IC1D) will always be in opposite states, so the output signal should always be HIGH. This would suggest that the circuit does nothing useful at all. To follow what is actually happening in this circuit, we must consider the logic states dynamically, rather than statically. No gate can respond instantly to any change in its input signals. Notice that input B to IC1D is the original input signal fed direct. But the signal that becomes "input A" to this gate must first pass through three other gate stages (IC1A, IC1B, and IC1C), each adding its own fraction of a second delay.

Let's assume we start out with a LOW external input signal. The output from each stage will be:

Gate	Inputs A	B	Stage Output
IC1A	1	0	1
IC1B	1	1	0
IC1C	1	0	1
IC1D	1	0	1

Now, when the external input signal switches from LOW to HIGH, we will get the following outputs:

Gate	Inputs A	B	Stage Output
IC1A	1	1	0
IC1B	1	0	1
IC1C	1	1	0
IC1D	0	1	1

The circuit's final output is the same as before. But, remember, the delay time added by each gate stage. When the external input first changes from LOW to HIGH, there will be a brief instant before the output of IC1C can respond and reverse its state. So, for a moment, "input A" to IC1D is still HIGH from the still unchanged IC1C output, but the direct "input B" signal has already gone HIGH, so the output of IC1D will be LOW until the circuit has a chance to stabilize to the new input state.

Similarly, when the external input signal goes from HIGH back to LOW, there will be a brief moment of reaction time, when the output of IC1C stays LOW, so IC1D sees identical inputs, and is LOW, until the circuit restabilizes, forcing the circuit's output HIGH again.

A typical set of input and output signals for this simple frequency-multiplier circuit are shown in Fig. 8-28. Notice that for every half-cycle of the input pulses, there is one complete output pulse cycle. In other words, there are two output pulses for every one input pulse. The output frequency is twice that of the input frequency.

Like any standard digital circuit, this frequency multiplier can only be used on rectangle wave input signals (or something closely resembling a rectangle wave). The duty cycle of the output signal probably will not be the same as the duty cycle of the input signal.

To achieve higher multiplication factors, you can use copies of this circuit in series. You can only achieve multiplication by factors of two in this manner. For exam-

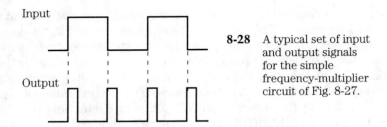

8-28 A typical set of input
and output signals
for the simple
frequency-multiplier
circuit of Fig. 8-27.

ple, the two-stage circuit shown in Fig. 8-29 multiplies the input frequency by four.
The three-stage-frequency-multiplier circuit illustrated in Fig. 8-30 has an output
frequency equal to eight times the frequency of the original input signal.

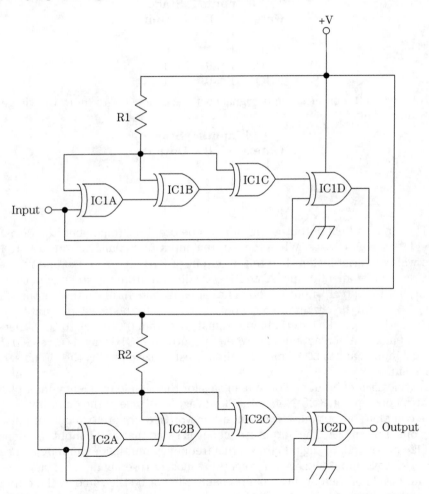

8-29 This two-stage frequency-multiplier circuit multiplies the input frequency
by four.

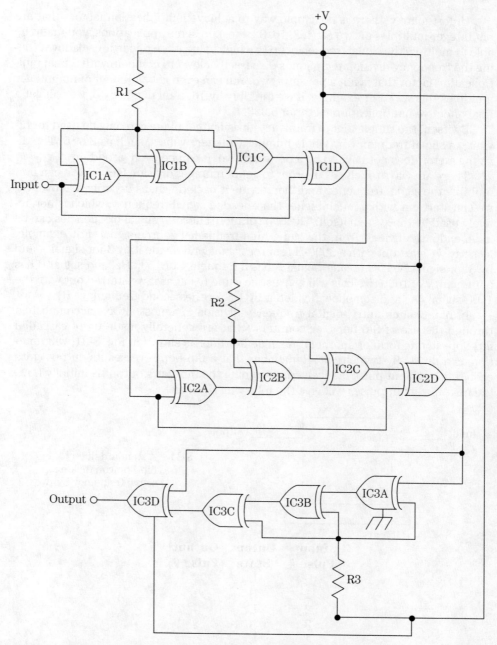

8-30 This three-stage frequency-multiplier circuit multiplies the input frequency by eight.

Unfortunately, there is no simple way to achieve multiplication factors that are not integer multiples of two (2, 4, 8, 16, 32, etc.). In some applications, you might be able to multiply the signal frequency up to a high value, then divide it back down (using the frequency division techniques discussed below) to come up with a final multiplication factor that is not a power of two. You can even get a number of noninteger multiples this way. For example, if we multiply by 16, then divide by 3, we will get a final effective multiplication factor of 5.333.

By itself, the simple digital frequency-division technique can only be used for dividing a signal frequency by whole number (integer) values, such as 2 or 5. The dividing factor does not have to be a power of two. It can be even or odd.

There are ways to "cheat" in order to get noninteger division factors. We can first multiply the signal frequency and then divide it back down, as described above. Or, we can start out with a much higher-than-needed signal frequency, which is not actively used by the system itself. The two (or more) desired signal frequencies can be independently divided from the same high-frequency source signal. For example, let's say we start out with a 3000-Hz source signal, and divide it by 2 for signal A and by 3 for signal B. This means, signal A has a frequency of 1500 Hz, and signal B has a frequency of 1000 Hz. This will give us the same result as if we started out with the 1500-signal (A) and somehow divided it by 1.5 for the second frequency (B).

In practical circuitry, digital frequency division is almost always accomplished through the use of flip-flops, or counters (that are generally made up of cascaded flip-flops in one form or another). A single flip-flop, as shown in Fig. 8-31, will function as a divide-by-two circuit. Remember that a flip-flop reverses its output state each time an input pulse is received. Assuming the flip-flop's output is initially LOW, the next few input pulses will give the following results:

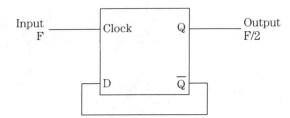

8-31 A standard digital flip-flop can divide a signal frequency by two.

Input Pulse #	Output State	Output Pulse #
0	L	0
1	H	1
2	L	1
3	H	2
4	L	2
5	H	3
6	L	3
7	H	4
8	L	4
9	H	5
10	L	5

and so forth.

Notice that for every two input pulse cycles, there is just one complete output pulse cycle. The output frequency is one-half the input frequency.

Once again, this circuit can only be used with rectangle-wave input signals (or something closely resembling a rectangle wave). The duty cycle of the output signal will probably not be the same as the duty cycle of the input signal.

By adding more flip-flops in series, we can divide by higher values. Each additional flip-flop raises the division factor by a power of two. It is easy enough to see why. Consider the three-stage divider circuit shown in Fig. 8-32. Flip-flop A divides the original input signal frequency (F) by two, so the output of this stage is $F/2$. Then flip-flop B divides the output frequency of flip-flop A by two, so this second stage has an output frequency of $F/4$. This frequency is again divided by 2 by flip-flop C, resulting in an output frequency of $F/8$. Theoretically, this can be continued indefinitely.

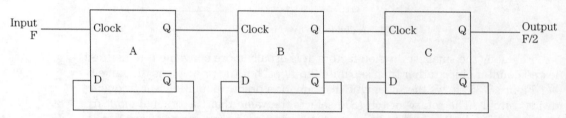

8-32   A three stage flip-flop circuit divides the input frequency by eight.

Cascading flip-flops in this manner creates a simple binary counter circuit. In effect, the circuit counts the input pulses. When the count has reached eight (the maximum count in this example circuit), the circuit produces an output pulse, then it starts over.

Suppose we need to divide the signal frequency by an integer that doesn't happen to be a nice, neat power of two? This is not the problem it would be in a frequency-multiplier circuit. In fact, the solution here is quite simple. All we have to do is to force the counter to reset itself prematurely when the desired value is reached. As examples, a divide-by-three circuit is shown in Fig. 8-33, and a divide-by-five circuit appears in Fig. 8-34.

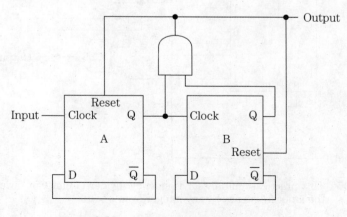

8-33   A divide-by-three circuit.

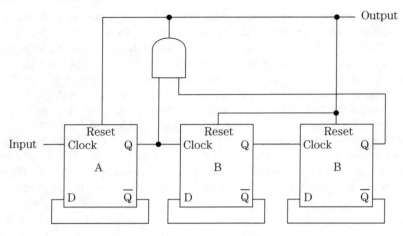

8-34 A divide-by-five circuit.

For all by the smallest division factors, it is usually more convenient to use dedicated counter chips rather than separate flip-flops, but the principle is the same.

Figure 8-35 shows the schematic diagram for a practical multidivisor frequency divider circuit. The rotary switch (S1) selects the value that the original input frequency will be divided by. This circuit permits division by any integer from 1 to 10. Yes, it might seem pointless to include divide-by-one, because the output frequency will simply be equal to the input frequency. But it doesn't add to the circuit's com-

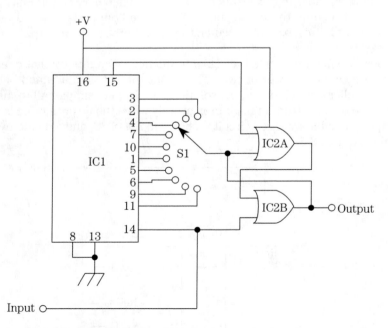

8-35 This practical multidivisor frequency divider circuit can divide the input frequency by any integer from 1 to 10.

plexity or cost, and it makes things easier if you should ever need the original undi-
vided frequency as an option. Besides, finding an SP10T rotary switch will probably
be simpler than a SP9T rotary switch, and a *divide-by-one* position is more practical than a dead-switch position.

It is easy to demonstrate the action of this circuit with a typical example. Let's assume that the input signal is a square wave with a frequency of 44 kHz (44,000 Hz).
The output frequencies for each position of switch S1 are summarized in Table 8-3.

**Table 8-3. This is a summary of typical outputs
for each switch setting in the multidivisor
frequency-divider circuit of Fig. 8-35**

Input frequency = 44,000 Hz	
Selected divisor	**Output frequency**
$\dfrac{F}{1}$	44,000 Hz
$\dfrac{F}{2}$	22,000 Hz
$\dfrac{F}{3}$	14,667 Hz
$\dfrac{F}{4}$	11,000 Hz
$\dfrac{F}{5}$	8,800 Hz
$\dfrac{F}{6}$	7,333 Hz
$\dfrac{F}{7}$	6,286 Hz
$\dfrac{F}{8}$	5,500 Hz
$\dfrac{F}{9}$	4,889 Hz
$\dfrac{F}{10}$	4,400 Hz

This circuit is very versatile, but surprisingly simple. Only three components are needed:

IC1   CD4017 decade counter
IC2   CD4001 quad NOR gate (two sections only)
S1   SP10T rotary switch

Only two of the four available sections in IC2 are used in this circuit. The remaining two sections can be used in other circuitry as part of a larger system, or they can be left idle. If any gates left unused, ground all of the unused inputs to avoid possible instability problems due to floating inputs. An unused gate with floating inputs can adversely affect the operation of other gates on the same semiconductor slab—that is, in the same IC package.

You might want to add line-filter capacitors between each IC's supply voltage pin and ground. This capacitor will help to protect the CMOS circuitry inside the IC from possible noise spikes in the supply voltage line. It should be mounted physically as close as possible to the body of the chip it is intended to protect. Each IC should have its own filter capacitor. The exact value of this capacitor is not at all critical and will not affect the operation of the circuit in any noticeable way. Typical values would be somewhere from 0.01 µF to 0.1 µF. Just use whatever capacitor in this range (or close to it) that you happen to have handy.

Synthesizing analog waveforms digitally

Ordinarily, digital circuitry can only be used with some form of rectangle wave. A digital circuit cannot normally accept any other analog waveform (such as a sine wave or sawtooth wave), as its input, nor can it generate any nonrectangle waveform as its output signal. This is because digital gates, by definition, can only recognize and operate on binary HIGH or LOW signals. They can't deal with any intermediate values, it always has to be one or the other.

But there are ways to get around this. This should be obvious from modern musical technology. Newer music synthesizers mostly use digital circuitry, and the popular CDs (compact discs) are a digital storage medium. Yet, both these systems are quite capable of creating (or reproducing) virtually any analog signal. It really isn't all that difficult to convert between the analog and digital realms, at least not on the theoretical level. The circuit that changes analog signals into strings of digital values is called an analog-to-digital, or A/D converter. For changing digital values into analog signals, a circuit known as a digital-to-analog, or D/A converter is used.

Unfortunately, it would not be appropriate in this book to go into detail about either A/D converter or D/A converter circuits. We will just briefly consider the basic principles involved, insofar as they apply directly to the subject of signal generation. If you are interested in the specific circuitry involved, such systems are covered extensively in many other books.

A digital system, by definition, works with numerical values. The binary number system used in digital electronics has just two possible values 0 (LOW) and 1 (HIGH). At first glance, that doesn't sound very useful. But what happens in our ordinary decimal number system when we need to express a value larger than 9, the largest available digit? We start a new column. So, 27 is two tens, and seven ones. Each new column is the next higher power of the system's base, which is ten in this case. So, for example:

$$41754 = \quad 4 \times 10^4 \quad 10 \times 10 \times 10 \times 10$$
$$+\ 1 \times 10^3 \quad 10 \times 10 \times 10$$
$$+\ 7 \times 10^2 \quad 10 \times 10$$
$$+\ 5 \times 10^1 \quad 10$$
$$+\ 4 \times 10^0 \quad 1$$

Any number raised to the zero power (X^0) equals 1.

The binary number system works in the same way, except we are dealing with powers of two, and a new column must be begun whenever the present column's value exceeds 1. So, for example:

$$1101001 = \quad 1 \times 2^6 = 1 \times 64 = 64$$
$$+\ 1 \times 2^5 = 1 \times 32 = 32$$
$$+\ 0 \times 2^4 = 0 \times 16 = 0$$
$$+\ 1 \times 2^3 = 1 \times 8 = 8$$
$$+\ 0 \times 2^2 = 0 \times 4 = 0$$
$$+\ 0 \times 2^1 = 0 \times 2 = 0$$
$$+\ 1 \times 2^0 = 1 \times 1 = 1$$

$$\text{(decimal equivalent)} \quad \overline{105}$$

For our purposes here, we don't really need to explore the binary number system any further than this. The key point is that combining multiple BInary digITs, or *bits* a digital circuit can express any desired numerical value.

In an A/D converter, the circuit repeatedly looks at the instantaneous amplitude of the analog input signal at various points throughout its cycle, as illustrated in Fig. 8-36. For each sample, the A/D converter circuit outputs a digital value that is the nearest to the present instantaneous signal level. Notice that the sampling rate in an A/D converter must be significantly higher than the signal frequency in order to digitally represent the waveshape with any accuracy. The sampling rate must be at least twice the signal frequency, or the system won't even be able to determine what the actual signal frequency is. This phenomena is known as *aliasing*.

A D/A converter, not surprisingly, works in just the opposite way. It accepts a multibit digital input value and puts out an analog signal with a voltage directly proportional to the input digital value. The result is varying step levels, each time the digital input value changes, as illustrated in Fig. 8-37.

Unfortunately, we still have a square, boxy signal, with lots of sharp corners—and therefore, lots of strong harmonics, that might not be part of the desired analog waveshape we want to synthesize (or reproduce). The solution is simple enough. A simple low-pass filter can round off the corners, creating a more analog-like waveform, as shown in Fig. 8-38. In many systems, this "smoothing" filter is nothing more than a simple relatively large-valued capacitor connected across the output of the D/A converter circuit.

The more steps per cycle, and the smaller each step is, the better a smooth analog waveform can be simulated digitally. This is illustrated in Fig. 8-39. Of course, there are always practical limits. After all, you can't very well design a digital system with an infinite number of infinitely small bits. Some level of compromise is always

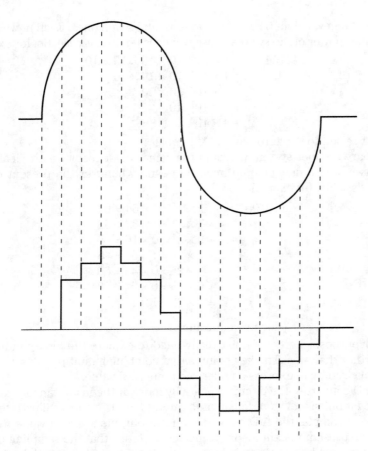

8-36 An A/D converter circuit repeatedly looks at the instantaneous amplitude of the analog input signal at various points throughout its cycle.

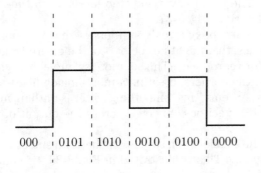

8-37 A D/A converter circuit produces an output signal comprised of varying step levels each time the digital input value changes.

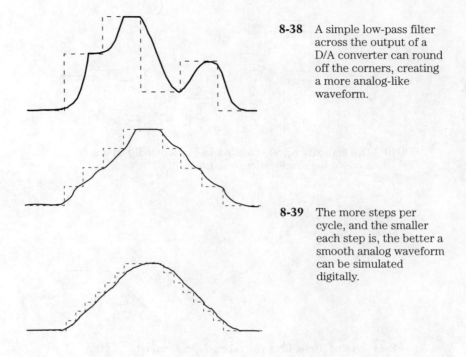

8-38 A simple low-pass filter across the output of a D/A converter can round off the corners, creating a more analog-like waveform.

8-39 The more steps per cycle, and the smaller each step is, the better a smooth analog waveform can be simulated digitally.

required. Still, with a well-designed circuit, most standard analog waveforms can be digitally simulated with remarkable accuracy. Remember, any practical analog oscillator or signal generator circuit will add some degree of distortion to the theoretically pure waveform. In digital synthesis, the distortion just takes a different form. In many practical systems, the digital distortion can be reduced to negligible levels, just as with good analog circuitry.

As an example of digital synthesis, we will aim for what is probably the most analog of all analog waveforms—the pure sine wave. Remember, a pure sine wave consists of the fundamental frequency, and no other frequency components at all. A sine wave's harmonic content is nominally zero. (We dealt with analog sine wave oscillator circuits back in chapter 2.)

The staircase waveform shown in Fig. 8-40 is as close as a digital system can come to the sinusodial waveshape by itself. The more steps there are building up to the peak levels, the better the approximation. A smoothing filter can take off the sharp edges, giving something reasonably close to a true sine wave, as illustrated in Fig. 8-41.

Many different circuits can be used to generate the necessary staircase wave. We can use one of the staircase-wave generator circuits discussed back in chapter 6. We can use some sort of computer/memory circuit that feeds the appropriate values out through a D/A converter. We can also use certain types of counter circuits. As it happens, this later choice turns out to result in the simplest and most direct circuitry, so that is the approach we will take here.

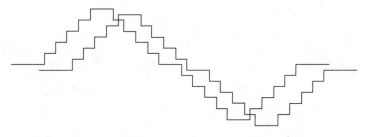

8-40 This staircase waveform is as close as a digital system can come to the sinusodial waveshape by itself.

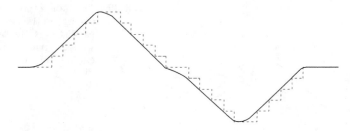

8-41 A smoothing filter can take off the sharp edges of the signal from Fig. 8-40, giving something reasonably close to a true sine wave.

We will be using a CD4018 counter IC. This device has five outputs. Each output is phase-shifted from the others by a delay of exactly one clock cycle apiece. This is easiest to see graphically, as shown in Fig. 8-42. If these outputs are summed together, we will get the desired staircase waveform, at least in theory. The basic circuit is illustrated in Fig. 8-43.

To get the correct output waveform, select the appropriate resistor values for R1 through R5. The obvious approach is to use equally valued resistors, but that would give an output signal like the one shown in Fig. 8-44. This signal is a lot closer to a triangle wave than a sine wave. Close, but no cigar yet.

You might want to breadboard this circuit and experiment with different resistor values. Monitor the output waveform with an oscilloscope and see how close you can come to a true sine wave. Perhaps you might want to do this before reading further.

You can generate lots of unusual and interesting waveshapes with this circuit by varying the resistor values. If you have experimented with the circuit as described above, you probably discovered that the pseudo-triangle wave that results from equal resistor values is pretty much as close as you can get to a sine wave. Is the task hopeless? No.

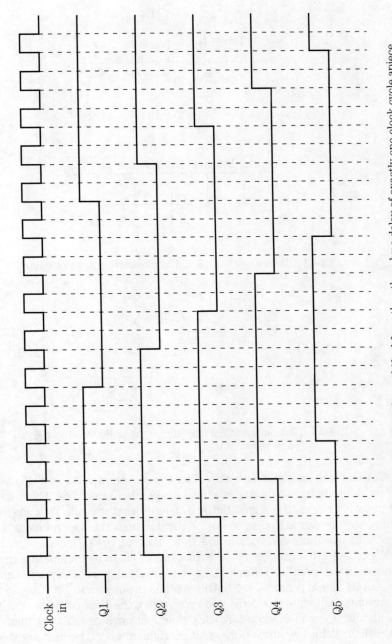

8-42   Each output of the CD4018 is phase-shifted from the others by a delay of exactly one clock cycle apiece.

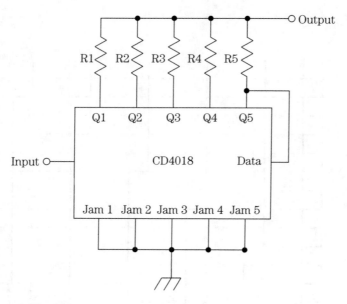

8-43   The basic CD4018 digital sine-wave generator circuit.

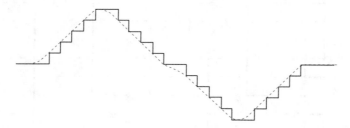

8-44   The output signal from the circuit of Fig. 8-43 with equal
value resistors.

Take a close look at a sine wave, like the one shown in Fig. 8-45. Notice that its
sloping sides change their angles at various points in the cycle. Close to the positive
and negative peaks, the signal flattens itself out a little bit. This suggests that the
highest step in our staircase should be wider than the ascending and descending
steps. This was indicated back in Fig. 8-40. Did you notice it?

We can get this wide *peak step* from our CD4018 by eliminating the Q5 output,
as shown in the modified circuit of Fig. 8-46. It is still part of the data generation pattern, but its signal is not added to the overall output signal.

Actually, equally-valued resistors won't give the best possible results, although
they should be good enough for many practical purposes. To calculate the exactly
best resistor values requires quite a bit of trigonometry. The best approach for most
electronics hobbyists would be to breadboard the circuit, and determine the best resistor values by trial and error.

The output frequency from this digital sine-wave synthesizer circuit is equal to
$\frac{1}{10}$ the input clock frequency used to drive the CD4018.

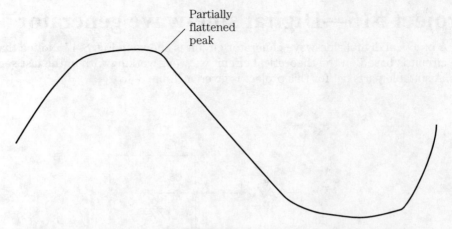

8-45   In a sine wave, the sloping sides change their angles at various points in the cycle.

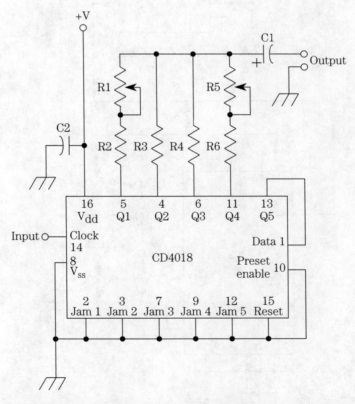

8-46   A better approximation of a sine wave can be achieved by modifying the circuit of Fig. 8-43 by eliminating the Q5 output.

Project #15—Digital sine-wave generator

A practical digital sine-wave generator circuit is shown in Fig. 8-47. Notice that this circuit is based on the theoretical circuit we were working with in the last section. A suitable parts list for this project is given in Table 8-4.

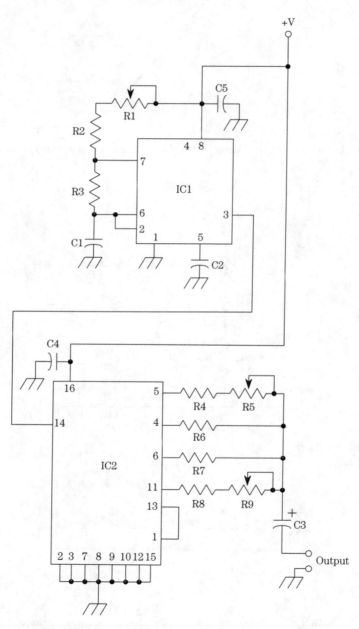

8-47   Project #15—Digital sine-wave generator.

**Table 8-4. Suggested parts list for Project #15
Digital sine-wave generator of Fig. 8-47**

IC1	7555 timer (see text)
IC2	CD4018
C1, C2	0.01-μF capacitor
C3	10-μF, 25-V electrolytic capacitor
C4, C5	0.1-μF capacitor
R1	1-MΩ potentiometer
R2	100-kΩ, ¼-W, 5% resistor
R3, R4, R8	1-kΩ, ¼-W, 5% resistor
R5, R9	5-kΩ trimpot
R6, R7	10-kΩ, ¼-W, 5% resistor (see text)

IC1 is a 7555 CMOS timer. A 555 timer chip can be substituted without making any changes in the circuitry. These two ICs are pin-for-pin compatible. This timer is wired as a basic astable multivibrator circuit, or a rectangle-wave generator. The output signal from this subcircuit serves as the clock input to the CD4018 (IC2).

Ideally, this clock signal should be a true square wave. Unfortunately that isn't possible using the 555 or 7555. But we can come fairly close, by making the value of the resistor (R3) between pins #7 and #6 as small as possible, as compared to the resistance between the supply voltage (V+) and pin #7 (R1 and R2 in series). For reliable operation, the minimum resistance value is about 1 kΩ (1000 ohms), so that is what we will use for R3. R1 is a potentiometer with a variable resistance from close to zero up to 1 MΩ (1,000,000 ohms), and it is in series with a 100-kΩ resistor (R2), so the total resistance varies from 100 kΩ up to 1.1 MΩ. This means the ratio of R_b (R_3) to R_a ($R_1 + R_2$) will always be at least 1:100. This will give us a clock signal that is reasonably close to a square wave. Our output pseudo-sine wave will not be significantly distorted by the clock's duty cycle.

Adjusting potentiometer R1 determines the clock frequency. The frequency formula, of course, is the familiar:

$$F = \frac{1}{\{0.693 C_1 (R_a + 2R_b)\}}$$

$$= \frac{1}{\{0.693 C_1 (R_1 + R_2 + 2R_3)\}}$$

At R1's minimum setting (nominally 0 ohms), the clock frequency works out to approximately:

$$F = \frac{1}{\{0.693 \times 0.00000001 \times (0 + 100000 + 2 \times 1000)\}}$$

$$= \frac{1}{\{0.00000000693 \times (100000 + 2000)\}}$$

$$= \frac{1}{(0.00000000693 \times 102000)}$$

$$= \frac{1}{0.000706}$$

$$= 1,416 \text{ Hz}$$

At R1's maximum setting (nominally 1,000,000 ohms), the clock frequency works out to about:

$$F = \frac{1}{\{0.693 \times 0.00000001 \times (1000000 + 100000 + 2 \times 1000)\}}$$

$$= \frac{1}{\{0.00000000693 \times (1100000 + 2000)\}}$$

$$= \frac{1}{(0.00000000693 \times 1102000)}$$

$$= \frac{1}{0.007636}$$

$$= 131 \text{ Hz}$$

The clock frequency can be adjusted from a little above 130 Hz to over 1.4 kHz. To form the pseudo-sine wave output signal, the CD4018 counts through a 10-step pattern, so the output frequency is equal to 10 times the clock frequency, for an output frequency range of about 1.3 kHz to 14 kHz. Capacitor C2 is included simply to ensure the stability of the timer.

I have already of calculated the resistor values in the counter portion of this project. However, for best results, the first and fourth resistances should be tunable so you can precisely calibrate the circuit for the best possible pseudo-sine wave at the output. It is best to perform this calibration while viewing the output waveshape on an oscilloscope. Alternately, if you have a pretty good ear, you can feed the output signal from this circuit through an audio amplifier and loudspeaker, and adjust these controls for the purest sounding tone. The calibration procedure can be a little tricky and time consuming until you get used to it, because these two controls interact somewhat. I recommend using screwdriver-adjustable trimpots instead of front-panel potentiometers for these two calibration controls. Once they have been correctly adjusted, they should be left alone.

For best results, I'd advise using an ohmmeter to check out the exact values of resistors R6 and R7 before wiring them into the circuit. Try several different resistors, and use the two that come closest to an actual measured value of 10,000 ohms. (5% tolerance resistors could range anywhere from 9500 ohms to 10,500 ohms.) If you don't want to bother making the ohmmeter measurements, you could use precision, low-tolerance resistors. But precision resistors are a lot more expensive than standard 5% tolerance resistors.

The exact value of capacitors C4 and C5 are not critical. Their function is to protect the CMOS ICs from any possible noise spikes on the supply voltage line. Each line filter capacitor should be mounted as physically close as possible to the body of the chip it is to protect. If a 555 timer is used instead of the 7555, capacitor C5 can be omitted from the circuit. It won't do any harm, but it won't serve any real purpose either.

Finally, capacitor C3 serves as a low-pass filter to round off the corners of the counter's output signal, and smooth it out to something that resembles a fairly decent sine wave.

Project #16—Precision digital clock

Figure 8-48 shows a useful digital clock circuit that generates very precise frequencies. A suitable parts list for this project appears as Table 8-5. As you can see, not many components are required for this project.

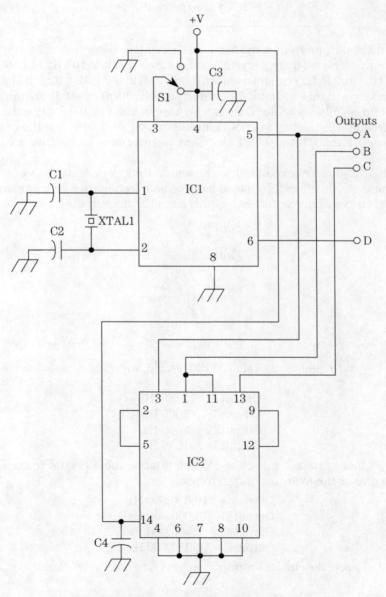

8-48   Project #16—Precision digital clock.

Table 8-5. Suggested parts list for Project #16
—Precision digital clock of Fig. 8-48

IC1	ICM7209 digital clock
IC2	CD4013
XTAL1	Crystal 10 kHz to 10 MHz (see text)
C1, C2	18-pF capacitor
C3, C4	0.1-μF capacitor
S1	SPDT switch (optional—see text)

The main output frequency from this circuit is determined by the crystal (XTAL1). Any standard quartz crystal cut for a frequency of 10 kHz (10,000 Hz) to 10 MHz (10,000,000 Hz) can be used with the ICM7209. Ideally, the crystal should have a load capacitance of 10 pF. But unfortunately, 30-pF crystals are much more common and readily available. They can be used in place of 10 pF crystals, but the circuit's frequency stability will not be quite as good. Similarly, use the best available capacitors for C1 and C2. Low-tolerance precision capacitors are recommended.

The ICM7209 offers two outputs, one with a frequency equal to ⅛ of the main clock frequency. IC2 offers a couple of intermediate frequencies for maximum flexibility. The relative frequency at each of this circuit's four outputs is:

$$\text{Output A} \quad F$$

$$\text{Output B} \quad \frac{F}{2}$$

$$\text{Output C} \quad \frac{F}{4}$$

$$\text{Output D} \quad \frac{F}{8}$$

If the lowest frequency crystal (10 kHz) is used, we will get the following output frequencies:

Output A	10,000 Hz
Output B	5000 Hz
Output C	2500 Hz
Output D	1250 Hz

At the other extreme, using the maximum acceptable crystal frequency (10 MHz), will give us the following output frequencies:

Output A	10,000,000 Hz
Output B	5,000,000 Hz
Output C	2,500,000 Hz
Output D	1,250,000 Hz

As you can see, despite its simplicity, this is quite a versatile circuit.

The ICM7209 consumes a minimum of power, but in many applications, it can still put a strain on the power supply when its signal isn't needed. Therefore, the chip includes a disable function. By grounding pin #3, the chip can be disabled, and will consume no power. There will be no output signal at this time, of course. This is the function of switch S1. It is optional, and can simply be omitted if the disable function is not needed in your intended application. Just wire pin #3 directly to the supply voltage (V+) line.

This clock circuit is not manually tunable, but the frequency can be changed by substituting a different crystal. In most practical applications, you will probably want to use a socket for the crystal, so it can be easily removed and replaced. In some applications, a rotary switch can be used to manually select from several different frequency crystals in parallel.

Index